# NATURAL ACTS

*A Sidelong View
of Science
and Nature*

# NATURAL ACTS

A *Sidelong View*
*of Science*
*and Nature*

## DAVID QUAMMEN

LYONS & BURFORD
PUBLISHERS

10 9 8 7 6 5 4 3 2    85 86 87 88

Copyright © 1985 by David Quammen

Library of Congress Cataloging in Publication Data
Quammen, David, 1948–
Natural acts.
"Nick Lyons books."
1. Natural history—Miscellanea.   I. Title.
QH45.5.Q36   1985      508      84-23613

MANUFACTURED IN THE UNITED STATES OF AMERICA

*Decorative art by Kris Ellingsen*

*The excerpt on pages 56-57 is reprinted from* The Fate of the Earth, *by Jonathan Schell. Copyright © 1982 by Jonathan Schell. Reprinted by permission of Alfred A. Knopf, Inc.*

*to M.E.Q. and W.A.Q.*
*who bought their boy a butterfly net*

# AUTHOR'S NOTE

Most of these pieces appeared first as instalments of the "Natural Acts" column that I write regularly for *Outside* magazine. The idea for a natural-science column in *Outside* was conceived originally by Janet Hopson, who wrote it with distinction for five years; Michael Rogers coined the column's title; I owe thanks to both of them.

The rest have appeared either in *Audubon, Esquire, Montana Outdoors,* or ("Desert Sanitaire") as a non-column feature in *Outside.*

Because the "Natural Acts" column always runs with a short title and a longer subtitle, I have tended to write it beneath title-subtitle combinations of my own choosing. But I see no point in forcing after-the-fact subtitles onto the other pieces. A subtitle in this book therefore indicates that the essay ran originally as a "Natural Acts" column.

I have also resisted the temptation to try to update every fact or number linked to a topical situation; those pieces that were *most* topical, filled with numbers that changed every time Congress voted or the Bureau of Reclamation revised a cost estimate, are simply not included in this book. In one case, for the sake of fairness, I've added a footnote. Otherwise it should be understood that a "now" or "at the present" alluded to in these essays might be any time between 1981 and 1984, and that contingent circumstances linked to those adverbials could be slightly—but not drastically—different when you read this.

I'm grateful to John Rasmus, Larry Burke, Gary Soucie, Les Line, Gene Stone, Dave Books, Kay Ellerhoff, Renée Wayne Golden, Nick

Lyons—and especially to David Schonauer of *Outside* and to my consulting biologist Kris Ellingsen—for editorial help and indulgence.

I am also quite grateful to Dr. Andy Sheldon, who taught me aquatic entomology at the time my life split fruitfully in two, and who told an anecdote about J. B. S. Haldane.

First publication of each of the pieces was as follows: "A Better Idea," *Outside* (November 1983); "Wool of Bat," *Outside* (August/September 1982); "The Widow Knows," *Outside* (April 1982); "The Troubled Gaze of the Octopus," *Outside* (July 1984); "Sympathy for the Devil," *Outside* (June/July 1981); "Has Success Spoiled the Crow?" *Outside* (October 1983); "Vox Populi," *Outside* (April 1983); "Rumors of a Snake," *Outside* (April 1984); "Avatars of the Soul in Malaya," *Outside* (March 1984); "A Republic of Cockroaches," *Outside* (May 1983); "The Excavation of Jack Horner," *Esquire* (scheduled December 1984); "The Lives of Eugène Marais," *Outside* (October 1981); "The Man with the Metal Nose," *Outside* (June 1983); "Voices in the Wilderness," *Outside* (July/August 1983); "Alias Benowitz Shoe Repair," *Outside* (December 1983); "The Tree People," *Outside* (January/February 1984); "Jeremy Bentham, the *Pietà*, and a Precious Few Grayling," *Audubon* (May 1982); "Homage to Bangi Bhale," *Outside* (May 1984); "Sanctuary," *Outside* (May 1982); "The Last Bison," *Outside* (November 1982); "Animal Rights and Beyond," *Outside* (June 1984); "The Big Goodbye," *Outside* (November 1981); "Love's Martyrs," *Outside* (September 1983); "Living Water," *Montana Outdoors* (May/June 1981); "A Deathly Chill," *Outside* (December/January 1983); "Is Sex Necessary?" *Outside* (October 1982); "Desert Sanitaire," *Outside* (February/March 1983); "The Miracle of Blubber," *Outside* (December/January 1982); "Dead Things in the Water," *Outside* (June/July 1982); "Yin and Yang in the Tularosa Basin," *Audubon* (scheduled January 1985).

# CONTENTS

---

# CAUSE FOR ALARUM

# ELOQUENT PRACTICES, NATURAL ACTS

# INTRODUCTION

---

## *"An Inordinate Fondness*
## *for Beetles"*

Biology has great potential as vulgar entertainment. For that matter so do geology, ecology, paleontology, and the history of astronomy. Browsing at the intricacies of the natural world and at the lives and works of the scientists who map that world can be fascinating, mesmeric, outrageous good fun. Alas, it can also be heart-squashingly boring. Edifying but deadly, like the novels of Henry James. Or just harmlessly quaint—nature as a vast curio shop with an inventory tending to cuteness. The crucial difference, at least to my biased view, is in the angle of approach. The choice of perspective. Lively writing about science and nature depends less on the offering of good answers, I think, than on the offering of good questions.

My own taste runs toward such as *What are the redeeming merits, if any, of the mosquito? Or Why is the act of sex invariably fatal for some species of salmon? Or Are crows too intelligent for their station in life? Why do certain bamboo species wait 120 years before bursting into bloom? How do seals stay cool in the Arctic? Does a termite colony constitute many little animals or one big one? Or perhaps best of all: Why are there so many different species of beetle?*

That last question has enthralled me for years—ever since I came across the delightful and precious fact that one of every four animals on Earth (by count of the number of species) is, yes, a beetle.

This is actually a conservative estimate, reflecting only the number of animal species that have been discovered and identified by science. Add up all the known species of mammals and birds and reptiles and amphibians, all the fish and crustaceans and protoplasmic tentacle-waving sea creatures, every brand of zooplankton that's ever been given a name, every type of worm, every flea every mite every spider, also of course every insect on the current entomological roster, and the total comes to around one-and-a-quarter million known species of animal. Of that vast assemblage, one in four is a beetle.

We're talking about an order called Coleoptera, containing 300,000 officially described species. And new beetles are being discovered almost every time some scientist waves a net through a rainforest. (A Smithsonian entomologist named Terry Erwin believes, from his study of jungle canopy in Peru, that there might be as many as *twelve million* species of beetle.)

Each of those 300,000 described species conforms to the basic Coleopteran pattern: an insect showing complete metamorphosis (progressing from egg to larva, then pupa, then adult) which in its adult form has biting mouthparts, a pair of front wings drastically modified into hard protective covers, a pair of lighter rear wings underneath those covers, and an extraordinarily strong cuticle over the whole body that looks like, and functions as, a suit of armor plating.

Within that basic pattern there is an unimaginable variety of shapes and colors and life strategies—vicious pinchers and rhinoceros horns on the head, anteater snouts, antennae like the most elaborate TV aerial, snapping hinges between thorax and abdomen that allow certain species to turn somersaults, light fixtures for signaling mates after dark, beetles as small as a sesame seed, beetles as large as a mouse, long scrawny beetles and husky broad-shouldered ones, leaf-eaters and fungus-eaters and meat-eaters, some that live underwater in rivers, some that burrow subway tunnels along the cambium layer of trees, some that gather and roll huge Sisyphean balls of dung. They are a very old as well as a very successful group of animals, dating back almost 250 million years, and in that stretch of eons they have had ample time to diversify. But cockroaches are equally old. So are dragonflies. So are sharks. So are lizards. Why then so cottonpickin' many species of coleopteran? Why 300,000 variations?

I don't know. I don't know of anyone who knows. I'm still waiting for some evolutionary biologist to propose a convincing explanation—but my secret hope is that no one can or will.

Meanwhile the great English geneticist J. B. S. Haldane has left us a valuable comment on this subject. Besides being an eminent scientist from a family of eminent scientists, Haldane was well known in the 1930s as a Marxist and a curmudgeon. Oral tradition among biologists records that Haldane was once cornered by a distinguished theologian. The theologian asked Haldane what inferences one could draw, from a study of the created world, as to the nature of its Creator. Haldane answered: "An inordinate fondness for beetles."

As it happens, J. B. S. Haldane also believed that science has great potential as vulgar entertainment. During the 1930s and 1940s he wrote a long series of science essays for the general public, most of which appeared in the *Daily Worker*. These short pieces had titles such as "Why I Admire Frogs," "Living in One's Skeleton," "Some Queer Beasts," and (following a case of theft from the British Museum, for which one entomologist went to prison) "Why Steal Beetles?" In spending his considerable wit and his precious working time to produce hundreds of popular essays, Haldane was perhaps the first in a tradition that is now burgeoning: the tradition of scientists who write graceful and accessible essays, on scientific subjects, for a lay readership.

Loren Eiseley continued that tradition, and today it is rich with the work of Stephen Jay Gould, Lewis Thomas, Freeman Dyson, Alan Lightman, Robert S. Desowitz and others. I would love to be able to claim a modest toehold in the same tradition. But I can't and I don't. Because I'm not a scientist.

What I am is a dilettante and a haunter of libraries and a snoop. The sort of person who has his nose in the way constantly during other people's field trips, asking too many foolish questions and occasionally scribbling notes. My own formal scientific training has been minuscule (and confined largely to the ecology of rivers). Gould and Thomas and Lightman actually *do* science, in addition to writing about it. I merely *follow* science. In my other set of pajamas I'm not a biologist but a novelist.

This autobiographical information is offered not because I imagine it has any inherent interest, but in a spirit of disclaimer, an effort at truth in packaging. The following is not a diet book nor a detective novel nor a collection of essays by a reputable scientist. Nor is it, for that matter, a string of straightforward dispatches from a "science reporter." It is the work of an outsider who is broadly curious but who can never remember the difference between meiosis and mitosis, who has nevertheless been invited to write on scientific subjects by a small number of charming but gullible magazine editors, who tries hard to keep the facts straight, who is not shy about offering opinions, and whose purpose in these pieces has been divided about equally between edification and vaudeville.

In the course of pondering what to say in this introduction, I invented an old saying that goes: "Put a magazine writer between hard covers, and immediately he thinks he's an essayist."

Of course I'm no exception. In defense of that claim I can say only that (1) the first section of this book, "All God's Vermin," was taking shape in my head as a sequence of essays long before I began making my living from magazine work, and that (2) I have tried to shape *all* of these pieces as essays more than as "features" or "profiles" or "articles," because old magazines go to the ragpicker, after all, whereas old books of essays are allowed to turn yellow with dignity on the shelf. (This doesn't apply to the piece here called "The Excavation of Jack Horner," which is clearly a profile, not an essay, commissioned by *Esquire* on those terms.) But please don't ask me to define *essay*, because the atmosphere could quickly grow ponderous. It's what Montaigne wrote. It's a filigreed editorial whose author doesn't know just which side he has argued until he reads the rough typescript. It's a small wobbly verbal dirtbike used for backcountry exploration, modest in horsepower yet under imperfect control of the cyclist. See what I mean about the atmosphere?

During the same time span from which these pieces come, I also wrote others that were definitely not essays—they were profiles or articles or reviews. Them you are being spared. The ragpicker has them already.

More important than categorizing the pieces of this book, though, is to say what unites them: subject matter and point of view. The subject matter is nature and the nature of science—with excursions into freshwater biology, geology, entomology, theoretical ecology, the history of astronomy, preservation issues, the role of bats in literature. The point of view is generally oblique and (or so I flatter myself) counterintuitive. My ambition has been to offer some small moments of constructive disorientation in the way nature is seen and thought about. Along the way I have been drawn in particular toward certain creatures that are conventionally judged repulsive, certain places that are conventionally judged desolate, certain humans and ideas that are conventionally judged crazy.

I have also found more fascination in the questions than in the answers. And here's a further question. On what grounds might we assume or hope that—despite the awesome puissance of modern science—any fertile mysteries still abide, unsolved, in the natural world? I can think of 300,000 reasons.

# ALL
# GOD'S
# VERMIN

# A
# BETTER IDEA

---

*Man and Cucumber
Cope As They Can*

Nature grants no monopolies in resourcefulness. She does not even seem to hold much with the notion of portioning it out hierarchically. Gold, she decrees, is where you find it.

Yet some biologists insist on glorifying us humans as the "most highly evolved" specimens of the animal world, far reaches in sophistication and ingenuity beyond such "lower" animals as the sponge, the barnacle, or that odd group of faceless geometrical sea creatures, the echinoderms. True, we have a fancy hand with an opposable thumb. True, we have an elaborate brain capable of memory, foresight, iambic pentameter, and malice. True, we know enough to come in out of the rain, usually. But not nearly so true is the anthropocentric presumption that *Homo sapiens* represents some sort of evolutionary culmination, embodying all the latest and best ideas. We are no such thing. And this humbling reality becomes especially clear with reference to a plump, homely gob of living matter known as the sea cucumber.

Lately the woman whose husband I am has been spying affectionately on these creatures, the sea cucumbers, along coral shelfs off the west coast of Mexico. She has returned with—besides an embarrassingly fine chestnut tan—one bizarre and edifying piece of information. It involves a biological problem which sea cucumbers and humans (in some parts of the world) happen to share: forcible entry by small parasitic fishes. What edifies is the ingenious method those poor stupid fat little cucumbers have devised for coping with this problem, compared to our own. You wouldn't believe it on a bet.

Sea cucumbers are not vegetables. They only look and act that way. In fact they are marine animals of the echinoderm phylum, a primitive group that also includes starfish, sea urchins, and two other star-shaped members called the feather-stars and the brittle-stars. Echinoderms are distinct from almost all other animal groups in being radially, rather than bilaterally, symmetrical. In other words they know top from bottom but not front from back nor left side from right. They all share a pentamerous anatomical organization, with most of their features occurring in fives: five axes of symmetry, five sets of each organ, five major arteries, and for those like the starfish and the brittle-stars, five legs. They have a mouth hidden under the belly, and an anus that generally marks the center of their back. The skin of an echinoderm is often described as "leathery" or "rubbery" but think instead of the texture of imperfectly cooked tripe. Imbedded in that skin are calcareous plates, in some cases quite small and with no inter-connections, constituting a minimal skeleton. Echinoderms have been known to stay in one spot, without moving, for up to two years. They have never heard of eyes. They developed all these eccentric proclivities, back in the Cambrian period a half billion years ago, before any consensus arose as to how an animal was supposed to behave. But just as the echinoderms are exceptional among animals, so the sea cucumbers are exceptional among echinoderms.

They retain the five-sided symmetry on the inside but don't give much hint of it externally. Sometime in the dim past they grew so tall and top-heavy that they have tipped over permanently onto one flank. The radial symmetry is now 90 degrees off kilter. Consequently they *do* have a discernible front: the end with the mouth, around which have been added a ring of tentacles like the leaf ends of celery. And they do have a rear: the end with the anus, to which we must address ourselves in a moment. They shuffle across the sea bottom in worm-like fashion, by means of muscular contractions and elongations that roll down their soft bodies in waves. Moving deliberately, they swallow the rich benthic mixture of sand and muck, strain the organic debris from it in their long simple gut, and pass the sterile sand out behind. Theoretically at least, they glide along like an open pipe while the sand, rippling faintly as it is cleaned, remains stationary.

In sea cucumbers (again, uniquely among all echinoderms) the skeletal plates are reduced to microscopic size and come in delicate patterns like snowflakes, but serve who knows what use. In overall body shape, some species resemble Italian sausages, some are more faithful to their garden namesake, some display the distinguished profile of a balloon overfilled precariously with tapioca. They range from the size of a baby gherkin to the size of a huge zucchini, one of those monstrous county-fair winners that gets its photo sent out on the AP wire. They are variously decorated in swirls and mottles and stripes of lavender, orange, yellow, parakeet green. Truly these guys are out in left field.

But it bothers them not. In the deepest trenches of the ocean they carry on blithely and quite successfully, working a zone that few other animals are equipped to explore. Researchers on the ocean abyss have discovered that, at a depth of 13,000 feet, sea cucumbers account for half of all the living organisms. Down at 28,000 feet, the sea cucumber majority rises to 90 percent. And at the ocean's bottommost bottom, 33,000 feet down in the Philippine Trench, almost no living creatures are to be found—except sea cucumbers.

In shallower waters, like those coral formations off the west coast of Mexico, they also get along well. This is in part because sea cucumbers have few natural predators, owing presumably to the various nasty poisons contained in the mucous secretions of their skin. Additionally, some species have developed the useful trick of self-mutilation: If a lobster or an otter or a snoopy human lays hold of one of this group, the sea cucumber constricts itself drastically at certain points along the body, and breaks into several pieces. The predator, ideally, will be satisfied with a middle or posterior section. All the sections are destined to die except the front end, with the mouth and tentacles. If this chunk is left in peace, from it will regenerate a new entire cucumber.

So it isn't their predators that pose the chief misery to these animals, it's the parasites that attack them from inside. One parasite in particular is shameless in the liberties taken, the indignities inflicted: a small, needle-thin species of cod called the pearlfish.

Pearlfish not only invade sea cucumbers but coolly set up housekeeping inside, feeding to some extent off the cucumber's internal organs, venturing back outside for additional forage, coming and going through the cucumber's anus. When the put-upon cucumber tightens its sphincter in an effort to deny entry, the pearlfish inserts first its pointed tail like a wedge, and then literally torques itself in through the clenched hole in a backward corkscrew motion. You'd think the little bugger was selling encyclopedias.

It was this dizzying bit of information—and a little more—that I learned from my personal consulting marine biologist. And I could sympathize wholeheartedly with the sea cucumber's position. Because, as it happened, I had lately been exposed to a similar threat myself.

In the tributaries of the Amazon River they call it the *candiru*, not a cod in this case but a parasitic catfish of the genus *Vandellia*, and it is known to attack humans. The candiru is needle-thin like the pearlfish, no more than two inches in length, with sharp teeth and an appetite for blood. It is also equipped with small spines angled rearward from the sides of its jaw, which serve like the barbs on an arrowhead or a fishhook: Once lodged in position within a host, the candiru has a very mean grip. Mainly it victimizes larger fish, whose gill openings the candiru enters, following scents of life upcurrent along the flow of expelled water as the victim fish breathes. But occasionally, in confusion, a stray candiru follows the life scents along a different flow: into some human so foolhardy as to urinate while bathing naked in candiru waters. For this delicate reason its English name, at least one of them, is the urethra fish.

On the jungle rivers of eastern Ecuador, draining the Andes toward the Amazon, common usage favors a still different name: the orofish. Very possibly this one derives from a bad pun. Every night for eight days, during a recent trip, I swam in the Rio Aguarico, but I was warned to beware of orofish. Use a good snug suit. An ingenious British companion suggested one other way we could guard against violation, while rinsing away Dr. Bronner's after a soaping on the beach: during submersion, hold the end.

Sea cucumbers don't have the luxury of any such precautions. They are routinely invaded by their own little nemesis, and routinely they must cope with the problem of an unwelcome visitor nibbling away at their viscera. For this the cucumbers have evolved a solution that is nothing less than wonderful in its directness, its dignified vehemence, its efficacy. They eviscerate.

They turn themselves inside out. They blast their own gut and internal organs out through their own anus, cast the whole smear away into the ocean, evicting in the process a single surprised pearlfish. And then—within as little as nine days, for some species—the sea cucumber regenerates a complete new internal anatomy.

Mere humans are not so deft. Once a candiru has made its painful entry, the problem is quite serious. An authority on the subject says: "Since the [candiru] have gill cover spines pointed to the rear, it is already too late once their presence is noticed since they cannot simply be pulled out. This has repeatedly caused death. If the afflicted individual does not want to have blood poisoning he must undergo an amputation." In which event, needless to say, there is no question of regeneration.

Now I ask you. Do not those lowly cucumbers have one on us?

# WOOL OF BAT

## Uses and Abuses
## of the World's Only Flying Mammal

From Pliny to Shakespeare to Tom McGuane, there has been a consensus: Any creature so grotesquely improbable as the common bat must perforce lend itself to some grotesquely improbable human use. The logic may be dubious but the notion is long-standing.

During the first century A.D. the elder Pliny, for instance, in his *Natural History*, suggested that a drop of bat's blood hidden under the pillow of a sleeping woman would serve as an aphrodisiac. Twelve centuries later Albertus Magnus claimed that smearing bat blood over your face like Coppertone would improve night vision. Macbeth's three witches, of course, are responsible for that famous recipe of which the partial ingredients are:

> *Eye of newt, and toe of frog,*
> *Wool of bat, and tongue of dog.*

An early colonial writer named John Lawson claimed that Indian children in North Carolina were often cured of a craving to eat dirt by feeding them roast bat on a skewer (though Lawson doesn't

8

say whether feeding dirt to adults might possibly cure their urge to roast a bat and make a child eat it). Rather more recently, McGuane's chortlesome second novel *The Bushwhacked Piano* describes the scheme of a certain C. J. Clovis, former fat man and one-legged con artist, to contract with mosquito-plagued municipalities for the erection of bat towers—each fully stocked with 1,500 bats dyed day-glo orange—for solving the local bug problem. The bats would devour the mosquitoes, theoretically, while the dye allowed citizens, of an evening, to watch their investment in action. And McGuane's C. J. Clovis is himself an echo of historical actuality: During the 1920s a Dr. Charles Campbell, of San Antonio, proposed "municipal bat roosts" to control mosquitoes and thereby malaria. At least one of those Campbell-style roosts was built, on Sugarloaf Key in Florida, by Mr. Richter C. Perky, who ran a fishing camp. Although accounts differ as to whether Perky ever stocked his tower with imported bats, or merely baited it to attract the native ones, orange day-glo dye can safely be ruled out. But wait. Even this isn't all.

Practical bats have just hit the news again. A late report in *American Heritage* magazine reveals that the United States government, during World War II, had a plan to use Mexican free-tailed bats for fire-bombing Japan.

The idea was to refrigerate these bats into hibernation, see, fit each with a small payload of napalm and a little bitty parachute, see, drop thousands like that from planes over Japanese cities, and hope for the best. No I'm not making this up. Your government. The research cost two million dollars.

Clearly the bat has captured human imaginations, and that may be because it seems triply oxymoronic: a flying mammal that sees in the dark by listening to its own silent screams. It is in truth an extraordinary animal, equipped with some startlingly sophisticated evolutionary adaptations, and represented around the world by a wild variety of different forms. If the bat is grotesquely improbable, so is Pablo Picasso.

Chiroptera is the collective name for this order of mammals, and the main defining characters are familiar: bald wings formed by thin flaps of skin stretched between hugely elongated fingers; the habit of feeding by night and resting by day; hind feet that lock closed automatically when suspending the body weight upside-down; and a system of precise echolocation, whereby the bat uses varying echoes from its own ultrasonic calls to find prey and steer itself through darkness. Anyone who has ever sat on a suburban porch while summer twilight faded has seen evidence of that last feature's uncanny efficiency. The erratic diving and swerving and swooping of those small black shapes reflects feeding success, not confusion. Their brains process the echo data at a rate unimaginable to us—though some researchers on the human brain have been eagerly studying how they do it—so that a cruising and squeaking bat, while avoiding an array of tricky obstacles, can catch one-third its own weight in flying insects within a half hour. Furthermore, they may have developed this sonar as much as 50 million years before we re-invented it.

The female bat possesses a single pair of mammaries, from which her young are tenderly nursed, and that fact among others led Linnaeus to suppose they were very closely related to humankind. Actually, bats seem to have evolved from some small earthbound insectivorous mammal, a common ancestor linking them with moles and shrews. We don't know for sure, since the earliest bat fossil (from Wyoming) is fully batlike, and no trace has been found of a transitional form.

But if the assumption about cousinhood is correct, it only highlights still more the uniqueness of bats, because they have gone—in complexity, in diversity, in longevity—so far beyond their relatives. Moles and shrews still feed almost exclusively on insects, while various bat species have diverged into diets of fruit, nectar and pollen, fish, other bats, small birds and rodents, lizards, and blood. Moles and shrews have remained restricted to specialized environments, while bats have dispersed across every tropical and temperate area of the planet. One genus of bats, comprising sixty species, is more widely distributed than any other genus of mammals, except perhaps *Homo*. Most striking, though, is the matter of age. A shrew in the wild can expect a lifespan of one year, under ideal conditions maybe two. Bats commonly live ten years and longer, in some cases twenty, and one record exists of a bat that survived to age twenty-four. They achieve this longevity thanks to large measures of sleep and hibernation. One informed guess is that a bat might spend five-sixths of its total life just hanging there, sound asleep.

Which seems fairly innocent. The bad reputation results, at. least in part, from a small group of South American bats called Desmodontidae, the vampires. Contrary to popular notion, these creatures do not grow as large as ravens, do not possess hollow fangs for sucking, do not usually victimize humans, and do not inhabit Transylvania, or any other part of Europe. Their deal is to make sneak attacks on Brazilian cattle, delicately nipping open a vein with sharp front incisors and then lapping away at the flow of blood with a dainty tongue. Also, rather oddly, they prefer to land a discreet distance from the intended cow and make their final approach by foot, back hunched up high, tiptoeing over the ground like some big-eared tarantula wearing a guilty smirk.

Here, I will concede, is a truly unsavory bat. Though perhaps even these Desmodontidae should get credit for a certain roguish charm. Anyhow, nobody—so far as history records—ever suggested training vampires to serve as official United States weapons of war. That distinction was reserved for *Tadarida brasiliensis*, the Mexican free-tail.

*T. brasiliensis* offered one major attraction over other potentially martial bats: abundance. There were 100 million of them roosting peaceably in just a few Texas caves.

Not even Tom McGuane and Albertus Magnus and Richter C. Perky all brainstorming over a bottle of George Dickel could have dreamed up an idea so demented as this napalm-bat thing: It took a dental surgeon from Pennsylvania named Lytle S. Adams. Seems that Dr. Adams was driving home from a vacation in New Mexico—where he had gazed wide-eyed at millions of *T. brasiliensis*, like one continuous pelt of lumpy brown fur, covering for acres the ceiling of the Carlsbad Caverns—when news of Pearl Harbor caught him upside the imagination. In first froth of patriotic outrage, and desirous of doing his bit, Adams thought of those bats. In less than two months, as the *American Heritage* article has it, Adams "somehow got the ear of President Franklin Roosevelt and convinced him that the idea warranted investigation." Under the circumstances "somehow" seems rather tantalizingly elliptic, but maybe FDR needed a little dental surgery and Dr. Adams pitched his idea before the gas had entirely worn off. Next he managed to interest an eminent Harvard chiroptologist (a bat expert, not a foot doctor) named Donald R. Griffin, and before long the National Defense Research Committee had signed on as sponsor. By now it was known as the "Adams Plan." Eventually the Army's Chemical Warfare Service, the NDRC, and the Navy (no reason submarines couldn't release these bats too) were all implicated in the buffoonery.

The first field tests were held at a remote airport in California on May 15, 1943. These were also, evidently, the last field tests. Adams and his colleagues discovered that *T. brasiliensis* could not always be put into hibernation, nor brought out of it, as promptly as might be convenient. And that the parachutes were a little too bitty. And that the incendiary capsules were a little too large. Groggy bats were tossed out of a plane. Many broke their wings. Some hit the ground without waking at all.

Yet there was poetic justice. A few other bats, armed on the ground with live napalm units, but spared the lethal jump, escaped from their handlers. These escapees flew off toward the nearest buildings—as indeed they were supposed to do, though preferably in Japan—which happened to be the airport hangars.

The hangars thereupon burned. So did a general's automobile. It did not seem auspicious to NDRC planners. The Adams Plan, in mercy to bats and chiroptophiles everywhere, was canceled.

Fair is foul, said the three witches, and foul is fair.

# THE
# WIDOW KNOWS

*Love and Death*
*Among* Latrodectus mactans

Here's a cheerful thought. Some knowledgeable people believe
that black widow spiders, like locusts and jack rabbits, come in
plagues.

Of course, no one has been keeping a running head count on
our total supply. Since the black widow is by nature shy, almost
fanatically discreet, the intermittent explosions of black widow
populations are best gauged by extrapolation from the number of
bites suffered by humans. At certain historical junctures of place
and time, those widow bites have reached what are called "epidemic" levels. Spain had such an epidemic in 1830. Eastern Russia
had another in 1838–39. France endured a peak forty years after
Russia, Uruguay in 1910, and then in 1926, the widows terrorized
Yugoslavia. These outbreaks of spider-bite and spider-fear were
all presumably caused by members of the genus *Latrodectus*, not
precisely the same species as our domestic black widow but closely
related. The last major episode of notoriety for *Latrodectus mactans*,
the American version, occurred a half century ago. In the autumn
of 1934, with Huey Long gamboling in the Senate and John
Dillinger freshly slain on the streets of Chicago, Americans were
suddenly worried about black widow spiders.

The widow boom that year was no doubt a combined result of climatic conditions, ecological cycles, and publicity. Through the mild winter and dry summer of 1933, more black widows had been surviving, raising larger broods, biting more humans, and getting more journalistic attention for doing so. The Associated Press carried continuing reports on the condition of two unfortunate men, one in Alabama and one in Idaho, both dangerously ill from widow bites. Local newspapers ran stories about barbarous mortal battles staged between black widows and scorpions, black widows and tarantulas, even black widows and snakes. With a peculiar repulsive fashionableness, black widows were *in*. Then *Scientific American* fanned the coals with an article announcing in careful detail why *Latrodectus mactans* should rightly be feared, and in November the sober journal *Science* published a note entitled "On the Great Abundance of the Black Widow Spider."

The note's author speculated about effects of the recent weather, commented that cousins and competitors of the widow seemed to be in eclipse, and concluded, "Possibly *L. mactans* is beginning to get the upper hand in the great struggle for existence." But the struggle against *whom?* To some nervous observers this upper-handhood seemed ominous in a creature with eight arms, not to mention a pair of poisonous chops, and so within months that statement from *Science* had been translated—now in the less sober pages of *Popular Mechanics*—into this one: "Aided by favorable climatic conditions, the spider has multiplied so rapidly that it is becoming a real menace to man." The headline in *Pop Mech* was "Wasps May End Black Widow Spider Menace." Entomologists were groping about for some spider-eating savior, a natural predator, a nemesis, to turn back those menacing surging hordes of *Latrodectus mactans*.

Two species of wasp were considered, a mud dauber that kills adult widows as food for its young, and another form that chews its way into the silken egg sac and deposits its own young where they will eat the unborn widows. There was also talk that *Latro-*

*dectus* might be subdued by a species of small parasitic fly, or by alligator lizards allowed to run rampant, or by a certain brave sort of toad, or perhaps by spider-eating spiders of the genus *Mimetus*. People were desperate.

Yet these scientists and other widow-watchers needn't have been quite so concerned.

The Great Widow Scare of 1934 came and went, like the Kingfish and Dillinger, but no airplanes were ever called out to spray. No frantic eradication campaign ever became necessary. *Latrodectus* continues to be widely distributed, present in every state and common in many, abundant in most of the warmer ones. Chances are good that, wherever you live, during the last year sometime you have sat down within ten feet of a black widow. (If you had occasion to use an old wooden outhouse, among the widow's favorite habitats, you might well have even been cheek by jowl with one.) Still, *Latrodectus* has never become a large-scale problem. Being bitten is nothing to take lightly, and if the victim is an old person or a small child, it can be fatal. But this is quite rare. The plague hasn't happened. The surging widow hordes, with their eight upper hands in the great struggle, have never come marching like driver ants over the hill. Why not? Partly because, in an ecological sense, the black widow knows its place.

Or, more specifically, *she* knows their place. The most efficient natural predator controlling any population of black widow spiders is, very likely, the female black widow. In that balance between vast reproductive potential and limited life-supporting resources—the Malthusian balance which is always delicate and always grim—it is evidently the female black widow who serves as chief gyroscope.

Concerning the behavior of the female *Latrodectus*, one canard should be clarified. A black widow spider is not like a mad dog or a rabid human: She does not kill except with good reason. She is not necrophilic or otherwise kinky. Yes, sometimes she does lasso the male after copulation, as he makes his break for the door, and suck him dry as a roasted chili. But that only happens if she is really hungry.

The female widow's hunger, or lack of it, is the standard against which certain life-or-death decisions are made, and those decisions exert a magnified geometric impact on overall *Latrodectus* population. When the concentration of widows in any area is high, competition is fierce for good web sites and for the local supply of food. As a result each widow female tends to be hungrier. The female has an amazing capacity to endure long stretches without any food—three or four months, and one widow in captivity went unfed for nine months, then was nursed back to health —which makes it unlikely the species could ever die out entirely from starvation. But as competition grows keen and food grows scarce, the female widow takes some drastic population-control measures that tip the balance back toward a state of lonely affluence. In the process she also slakes her own immediate hunger. The first of these measures involves mating.

The male widow, much smaller than the female and more mobile across unfamiliar terrain, takes the initiative. He appears at the edge of the female's web, sets his feet on a few strands and, by bobbing his abdomen, causes the whole network to vibrate. This is the mating offer. If the female is not in condition to breed, or not in the mood, she will not respond. But if she happens also to be very hungry, she may wait cryptically for him to venture within reach, then grab him and swathe him in silk and eat him. If she *is* ready to mate, there is an answering pattern of vibrations from her. After about two hours of foreplay, during which he wraps her in a loose veil of silk, they copulate. That takes five minutes, but the logistics are exotic.

As they uncouple, the question again arises: Is she hungry? If so, he is snatched back and devoured. If not, he is allowed to leave peaceably, or he may linger in a corner of her web, safely ensconced as mate emeritus, until accelerated senescence kills him a few days later. Whichever of these is the outcome, the female has remained in character, neither romantic nor sadistic but merely practical, all eight of her eyes fixed on the basics of survival. She is the embodied future of her species, far more so than the male, and in a dim instinctive sense she knows it.

The same sort of cool pragmatism governs her maternal behavior: She is a solicitous mother, but only up to a definite point. Her eggs are laid—anywhere from 25 to 1,000 in a batch, on average about 200—and then wrapped by her in a silken cocoon. This egg sac is watertight, shields the eggs from sunlight and predators, and helps keep them warmly incubated. The mother has taken great care over it. She will fight to protect it. She will spend immense effort moving it from one part of the web to another, sunlight to shade, for the sake of maintaining its optimal temperature. Then when the eggs hatch and her young crawl out, if the message of hunger has by now again reached her and the prospects seem meager for black widows, she will eat them.

Or at least some of them. After thirty or forty she may stop, and the rest of the spiderlings will go off to face all the other obstacles, including starvation, between them and adulthood.

Does the maternal widow have some instinctive awareness about how many young spiders—given the circumstances of food supply at the moment of hatching, which may be quite different from those when she laid the eggs—have a chance to survive? We don't know. Is a male widow sexually capable of servicing more than one female? Is the possibility of his fathering further broods part of what influences his first mate, in the matter of whether she will kill him or let him escape? We don't know. And the female, with her edgy perceptions of hunger and her hair-trigger readiness to cannibalize close relatives—has she been ingeniously programmed by evolution to serve as a ruthless and economical natural control on fluctuations in the population of her species? It seems plausible. But we don't really know.

Plenty has been written about red hourglasses and poisonous bites; precious little about the biology and population dynamics of the black widow. Who *does* know what graceful intricacies lurk in the brain of *Latrodectus?*

# THE
# TROUBLED GAZE
# OF THE OCTOPUS

## Good Eyesight and Mental Hygiene
## on the Ocean Floor

In *Gravity's Rainbow*, Thomas Pynchon's great steaming slag-heap of a novel, there is a memorable scene in which an enormous octopus comes slouching out of the sea, grabs a young woman around the waist with one sucker-studded arm, and tries to drag her away back into the water. Echoes of King Kong and Fay Wray. The woman is rescued by Pynchon's hero, a certain Tyrone Slothrop, who pummels the octopus over the head with a wine bottle (to no effect), and then distracts it by offering a tasty crab. "In their brief time together Slothrop forms the impression that this octopus is not in good mental health," Pynchon tells us. I have sometimes had the same feeling about Thomas Pynchon, but never mind. The point of recalling that octopus scene, in particular, is that I for one always took it to be luridly and outlandishly implausible, an excursion to the outback of the surreal, another hallucinatory cartoon caper like so much else that comes out of Mr. Pynchon's febrile imagination.

Turns out, though, that I was unduly skeptical. Turns out that this sequence might contain more truth than poetry. The details of anatomy and scale and behavior have a sound basis in science; only the matter of motivation remains unsettled. The species in question is *Octopus dofleini*, otherwise known as the giant Pacific octopus.

Does this creature really lay hold, in such peremptory manner, of unsuspecting human beach-goers? And if so, just what has it got in mind?

The concept of *mind* is not inappropriate as applied to the octopi, since these creatures have by far the most highly developed brain in their province of the animal kingdom. They belong to the phylum Mollusca, a large group of invertebrates mainly characterized by soft bodies, hard shells, and rather primitive patterns of anatomical organization, well suited to surviving inconspicuously on the sea bottom. Typical of the Mollusca are clams, oysters, snails; the octopi (and to a lesser extent their near relatives, the squids) are decidedly untypical. They are an evolutionary anomaly, a class of genius misfits who have advanced far beyond their origins.

The octopi have an elaborate fourteen-lobed brain, an organ so large that their brain-to-body-weight ratio exceeds that of most fish and reptiles. Mentally, they are more on a level with birds and mammals. They possess a capacity for learning, memory and considered behavior that makes them—with the exception of marine mammals—the most intelligent of all sea-dwelling animals. In a laboratory, they tend to be good at mazes, and perform well in tests of discrimination among visual symbols. This last talent depends partly upon their acute eyesight. Every octopus looks out at the world through a pair of extraordinary eyes—eyes about which, to a human, there is something unexpectedly and disquietingly familiar.

"The animal has eyes that stare back," according to Martin J. Wells, a British zoologist who is one of the world's experts in octopus physiology and behavior. "It responds to movement, cowering if anything large approaches it, or leaning forward in an alert and interested manner to examine small happenings in its visual field." Jacques Cousteau goes a bit farther: "When a diver sees a giant octopus in the dim water, its great eyes fixed on him, he feels a strange sensation of respect, as though he were in the presence of a very wise and very old animal, whose tranquility it would be best not to disturb." One of Cousteau's assistants adds: "I have often had the impression that they are 'reflecting.' " Other divers and lab researchers make the same sort of comment, describing the same eerie sense of encounter, recognition, even mutuality. Lately I've had occasion to experience it myself, during three evenings of octopus-watching in a small university room filled with quietly gurgling tanks: the potent, expressive gaze of the octopus. These animals don't just gape at you glassily, like a walleye. They make eye contact, as though they are someone you should know.

One of the reasons for the potency of that stare is simply a matter of proportion. Relative to the body size of a given octopus, the eyes are, like the brain, unusually large. (The ultimate record in this regard belongs to that octopus cousin, the giant squid— with an eyeball up to fifteen inches across, largest on Earth and twice the size of the eye of a blue whale.) Octopus eyes are also protrusive and mobile, bulging up periscopically when the crea-ture's attention is caught, swiveling far enough fore and aft to cover all 360 degrees of horizon. But the real magic behind the octopod gaze is that those eyes bear a startling structural similarity to our own.

It's an exemplary instance of the phenomenon called *convergent evolution*. Two separate evolutionary paths are followed for mil-lions of years by two disparate groups of creatures, arriving even-tually at two separate but (coincidentally) very similar solutions to a common problem. In this case, the problem of translating incident light rays into coherent images conveyable to the brain. The vertebrate eye—the model we humans share with cougars and eagles and rattlesnakes, all having inherited the pattern com-monly—is an ingenious contrivance combining a cornea, a crys-talline lens, an adjustable iris, and a retina. That such an organ evolved even once, within the vertebrate line, represents a miraculous triumph of time and trial-and-error over improbability. The still larger miracle is that *two* very similar versions have appeared independently. The other belongs exclusively to the octopi and their close kin. Each of those squid and octopus eyes consists of a cornea, a crystalline lens, an adjustable iris, and a retina, functioning together in much the same way as ours.

And the octopi are also endowed with an eyelid, so that they can wink at us fraternally.

Among all the sea's vigilant octopod eyeballs, the most imposing belong to *Octopus dofleini*. This is the giant that has impressed Cousteau and others with its dignified presence, as though it were "a very wise and very old animal." The wisdom part is quite possible, but the great agedness is an illusion.

Octopi grow quickly to adulthood and die at an early age, in most cases right after their first breeding experience. Two or three years is a full lifespan, even for the larger Mediterranean octopi that might grow in that brief time from the size of a flea to the size of a collie. An octopus can achieve such speedy growth— almost doubling its weight each month throughout most of its life—because of its exceptional metabolic efficiency in converting food protein to octopus protein. And at this process, *O. dofleini* is probably unsurpassed. Surviving longer than other species, to the grand age of five or maybe a little beyond, some giant Pacific octopi attain awesome sizes.

They live mainly in sea-bottom caves along the coast of the northern Pacific, from California up through British Columbia and Alaska, and across to Japan. The caves give them security from predation and, during mating time, a good place to brood eggs. They seem to prefer that range of moderate depths from the intertidal zone down to 100 fathoms, and by most accounts they are exceedingly shy. Until three or four decades ago, *dofleini* were known almost solely from commercial fishermen, who occasionally, inadvertently, and probably to their own vast alarm, brought one up in a trawl net. After World War II, improved diving equipment (especially scuba) opened a new degree of human access to those deeper caves. Reports of larger and larger *dofleini* began to appear. Cousteau mentions one specimen, spotted off Seattle, with an arm-span of thirty feet and a weight of around 200 pounds. Another diver has told William High, of the National Marine Fisheries Service, about bringing up several 400-pound *dofleini* during his commercial octopus-fishing days, as well as a single huge individual that went 600 pounds.

Then the inevitable happened. Someone had a clever idea, and in 1956 hundreds of divers converged on Puget Sound, to compete in an event billed as the World Octopus-Wrestling Championship. It became an annual tradition.

The biggest specimens of *dofleini* were smoked out of their caves with solutions of noxious chemicals, and wrestled up onto land by divers working in teams, there to be weighed and measured and admired. Dealing underwater with an octopus that large requires—from a human in scuba gear—equal measures of skill, cool-headedness, and lunatic daring. *Dofleini* are naturally timid but also quite strong, and they do have four times as many arms. Panic-stricken *dofleini* have been known to pin a man's arms to his sides, pull off his face mask, yank out his mouthpiece. There have even been several instances when a big octopus pounced on the back of an unsuspecting diver, from a rock ledge overhead. If one diver becomes trapped, in a situation like that, the usual procedure seems to be for his buddy to commence slicing away octopod arms with a knife, or to go for the eye with a spear.

After those "wrestling" championships each year, the healthy octopi were carefully released back into the sea. No harm done. No permanent toll on the giant octopus population. Right? At least that was the assumption. But one woman diver who took part in the round-ups has told Cousteau: "It is hard to keep in mind that octopuses of the size and weight of these are really very fragile animals, highly developed and with a very sensitive nervous system. They seem to succumb easily to nervous disorders. If a diver is too rough with an octopus, even without actually hurting it physically, it happens that the animal goes into a state of emotional shock and sometimes dies."

\*        \*        \*

Let me recapitulate. It seems that (1) octopi in the 200-pound range, or larger, cowering in sea caves off the coast near Seattle, and (2) known for their high-strung susceptibility to nervous disorders, have been (3) kidnapped and terrorized intermittently by strange visitors in black neoprene, these latter often armed with knives and spears. Consequently, it can be assumed that (4) eyeballs human-like but the size of grapefruit now gaze out from those caves, furtively, trepidatiously, some of them no doubt looking just a bit addled. Looking as though they might belong to animals that, like Pynchon's beast, are in not quite the best mental health.

My own view is this: Any giant octopus that grabs hold of a passing human probably has some pretty good reason. If not an unanswerable grievance, then at least a plausible insanity defense.

Or maybe the creature is just desperate to communicate. Snatching that lone human up by the rubber lapels. Exigent as the Ancient Mariner. Transfixing the man or the woman with a big glittering eye. *Listen. We know who you are. And we've seen what you do.* But unfortunately the octopi, for all their intelligence, for all their sensitivity, for all their remarkable evolutionary sophistication, are born mute.

# SYMPATHY
# FOR THE DEVIL

---

*A More Generous View
of the World's Most Despised Animal*

Undeniably they have a lot to answer for: malaria, yellow fever, dengue, encephalitis, filariasis, and the ominous tiny whine that begins homing around your ear just after you've gotten comfortable in the sleeping bag. All these griefs and others are the handiwork of that perfidious family of biting flies known as Culicidae, the mosquitoes. They assist in the murder of millions of humans each year, carry ghastly illness to millions more, and drive not a few of the rest of us temporarily insane. They are out for blood.

Mosquitoes have been around for 50 million years, which has given them time to figure all the angles. Judged either by sheer numbers, or by the scope of their worldwide distribution, or by their resistance to enemies and natural catastrophe, they are one of the great success stories on the planet. They come in 2,700 different species. They inhabit almost every land surface, from Arctic tundra to downtown London to equatorial Brazil, from the Sahara to the Himalaya, though best of all they like tropical rainforests, where three-quarters of their species lurk. Mosquitoes and rainforests, in fact, go together like gigolos and bridge tournaments, insurance salesmen and Elks lunches, panhandlers and. . . . But more on all that in a moment.

They hatch and grow to maturity in water, any entrapment of quiet water, however transient or funky. A soggy latrine, for instance, suits them fine. The still edge of a crystalline stream is fine. In the flooded footprint of an elephant, you might find a hundred mosquitoes. As innocent youngsters they use facial bristles resembling cranberry rakes to comb these waters for smorgasbord, but on attaining adulthood, they are out for blood.

It isn't a necessity for individual survival, quenching that blood thirst, just a prerequisite of motherhood. Male mosquitoes do not even bite. A male mosquito lives his short, gentle adult life content, like a swallowtail butterfly, to sip nectar from flowers. As with black widow spiders and mantids, it is only the female that is fearsome. Make of that what larger lessons you dare.

She relies on the blood of vertebrates—mainly warm-blooded ones but also sometimes reptiles and frogs—to finance, metabolically, the development of her eggs.

A female mosquito in a full lifetime will lay about ten separate batches of eggs, roughly 200 in a batch. That's a large order of embryonic tissue to be manufactured in one wispy body, and to manage it the female needs a rich source of protein; the sugary juice of flowers will deliver quick energy to wing muscles, but it won't help her build 2,000 new bodies. So she has evolved a hypodermic proboscis and learned how to *steal* protein in one of its richest forms, hemoglobin. In some species her first brood will develop before she has tasted blood, but after that she must have a bellyfull for each set of eggs coming to term.

When she drinks, she drinks deeply: The average blood meal amounts to 2½ times the original weight of the insect. Picture Audrey Hepburn sitting down to a steak dinner, getting up from the table weighing 380 pounds; then, for that matter, flying away. In the Canadian Arctic, where species of the genus *Aedes* emerge in savage, sky-darkening swarms like nothing seen even in the Amazon, and work under pressure of time because of the short summer season, an unprotected human could be bitten 9,000 times per minute. At that rate a large man would lose half his total blood in two hours. Arctic hares and reindeer move to higher ground, or die. And sometimes solid mats of *Aedes* will continue sucking the cool blood from a carcass.

Evidently the female tracks her way to a blood donor by flying upwind toward a source of warmer air, or air that is both warm and moist, or that contains an excess of carbon dioxide, or a combination of all three. The experts aren't sure. Perspiration, involving both higher skin temperature and released moisture, is one good way to attract her attention. In Italy it is established folk wisdom that to sleep with a pig in the bedroom is to protect oneself from malaria, presumably because the pig, operating at a higher body temperature, will be preferred by mosquitoes. And at the turn of the century, Professor Giovanni Grassi, then Italy's foremost zoologist, pointed out that garrulous people seemed to be bitten more often than those who kept their mouths shut. The experts aren't sure, but the Italians are full of ideas.

Guided by $CO_2$ or idle chatter or distaste for pork or whatever, a female mosquito lands on the earlobe of a human, drives her proboscis (actually a thin bundle of tools that includes two tubular stylets for carrying fluid and four serrated ones for cutting) through the skin, gropes with it until she taps a capillary, and then an elaborate interaction begins. Her saliva flows down one tube into the wound, retarding coagulation of the spilled blood and provoking an allergic reaction that will later be symptomized by itching. A suction pump in her head draws blood up the other tube, a valve closes, another pump pulls the blood back into her gut. And that alternate pumping and valving continues quickly for three orgiastic minutes, until her abdomen is stretched full like a great bloody balloon, or until a fast human hand ends her maternal career, whichever comes first.

But in the meantime, if she is an *Anopheles gambiae* in Nigeria, the protozoa that cause malaria may be streaming into the wound with her saliva, heading immediately off to set up bivouac in the human's liver. Or if she is *Aedes aegypti* in Costa Rica, she may be drooling out an advance phalanx of the yellow fever virus. If she is *Culex pipiens* in Malaysia, long tiny larvae of the filaria worm may be squirting from her snout like a stage magician's spring-work snakes, dispersing to breed in the unfortunate person's lymph nodes and eventually clog them, causing elephantiasis. Definitely, this is misanthropic behavior.

No wonder, then, that in the rogues' pantheon of those select creatures not only noxious in their essential character but fur-thermore lacking any imaginable forgiving graces, the mosquito is generally ranked beyond even the wood tick, the wolverine, or the black toy poodle. The mosquito, says common bias and on this the experts agree, is an unmitigated pain in the ass.

But I don't see it that way. To begin with, the family is not monolithic, and it does have—from even the human perspective—its beneficent representatives. In northern Canada, for instance, *Aedes nigripes* is an important pollinator of arctic orchids. In Ethio-pia, *Toxorhynchites brevipalpis* as a larva preys voraciously on the larvae of other mosquitoes, malaria carriers, and then *Toxo* itself transforms to a lovely huge iridescent adult that, male or female, drinks only plant juices and would not dream of biting a human.

But even discounting these aberrations, and judging it by only its most notorious infamies, the mosquito is taking a bad rap. It has been victimized, I submit to you, by a strong case of an-thropocentric press-agentry. In fact the little sucker can be viewed, with only a small bit of squinting, as one of the great ecological heroes of planet Earth. If you consider rainforest preservation.

The chief point of blame, with mosquitoes, happens also to be the chief point of merit: They make tropical rainforests, for humans, virtually uninhabitable.

Tropical rainforest constitutes by far the world's richest and most complex ecosystem, a boggling entanglement of life forms and habits and equilibriums and relationships. Those equatorial forests—mainly confined to the Amazon, the Congo basin and Southeast Asia—account for only a small fraction of the Earth's surface, but serve as home for roughly *half* of the Earth's total plant and animal species, including 2,000 kinds of mosquito. But rainforests lately, in case you haven't heard, are under siege.

They are being clear-cut for cattle ranching, mowed down with bulldozers and pulped for paper, corded into firewood, gobbled up hourly by human development on the march. The current rate of loss amounts to eight acres of rainforest gone poof since you began reading this sentence; within a generation, at that pace, the Amazon will look like New Jersey. Conservation groups are raising a clamor, a few of the equatorial governments are adopting plans for marginal preservation. But no one and no thing has done more to delay this catastrophe, over the past 10,000 years, than the mosquito.

The great episode of ecological disequilibrium we call "human history" began, so the Leakey family tell us, in equatorial Africa. Then immediately the focus of intensity shifted elsewhere. What deterred mankind, at least until this half of this century, from hacking space for his farms and his cities out of the tropical forests? Yellow fever did, and malaria, dengue, filariasis, o'nyong-nyong fever.

Clear the vegetation from the brink of a jungle waterhole, move in with tents and cattle and Jeeps, and *Anopheles gambiae*, not normally native there, will arrive within a month, bringing malaria. Cut the tall timber from five acres of rainforest, and species of infectious *Aedes*—which would otherwise live out their lives in the high forest canopy, passing yellow fever between monkeys—will literally fall on you, and begin biting before your chainsaw has cooled. Nurturing not only more species of snake and bird than anywhere else on earth, but also more forms of disease-causing microbe, and more mosquitoes to carry them, tropical forests are elaborately booby-trapped against disruption.

The native forest peoples gradually acquired some immunity to these diseases, and their nondisruptive hunting-and-gathering economies minimized their exposure to mosquitoes that favored the canopy or disturbed ground. Meanwhile the occasional white interlopers, the agents of empire, remained vulnerable. West Africa in high colonial days became known as "the white man's grave."

So as Europe was being stripped of its virgin woods, and India and China, and the North American heartland, the rainforests escaped, lasting into the late twentieth century—with some chance at least that they may endure a bit longer. Thanks to what? Thanks to ten million generations of jungle-loving, disease-bearing, blood-sucking insects: the Culicidae, nature's Viet Cong.

And a time, says Ecclesiastes, to every purpose.

# HAS SUCCESS
# SPOILED THE CROW?

---

*The Puzzling Case File
on the World's Smartest Bird*

Any person with no steady job and no children naturally finds
time for a sizable amount of utterly idle speculation. For instance,
me—I've developed a theory about crows. It goes like this:

Crows are bored. They suffer from being too intelligent for
their station in life. Respectable evolutionary success is simply
not, for these brainy and complex birds, enough. They are dissatisfied with the narrow goals and horizons of that tired old
Darwinian struggle. On the lookout for a new challenge. See them
there, lined up conspiratorially along a fence rail or a high wire,
shoulder to shoulder, alert, self-contained, missing nothing. Feeling discreetly thwarted. Waiting, like an ambitious understudy,
for their break. Dolphins and whales and chimpanzees get all the
fawning publicity, great fuss made over their near-human intelligence. But don't be fooled. Crows are not stupid. Far from it.
They are merely underachievers. They are bored.

Most likely it runs in their genes, along with the black plumage
and the talent for vocal mimicry. Crows belong to a remarkable
family of birds known as the Corvidae, also including ravens,
magpies, jackdaws and jays, and the case file on this entire clan
is so full of prodigious and quirky behavior that it cries out for

interpretation not by an ornithologist but a psychiatrist. Or, failing that, some ignoramus with a supple theory. Computerized ecologists can give us those fancy equations depicting the whole course of a creature's life history in terms of energy allotment to every physical need, with variables for fertility and senility and hunger and motherly love; but they haven't yet programmed in a variable for boredom. No wonder the Corvidae dossier is still packed with unanswered questions.

At first glance, though, all is normal: Crows and their corvid relatives seem to lead an exemplary birdlike existence. The home life is stable and protective. Monogamy is the rule, and most mated pairs stay together until death. Courtship is elaborate, even rather tender, with the male doing a good bit of bowing and dancing and jiving, not to mention supplying his intended with food; eventually he offers the first scrap of nesting material as a sly hint that they get on with it. While she incubates a clutch of four to six eggs, he continues to furnish the groceries, and stands watch nearby at night. Then for a month after hatching, both parents dote on the young. Despite strenuous care, mortality among fledglings is routinely high, sometimes as high as 70 percent, but all this crib death is counterbalanced by the longevity of the adults. Twenty-year-old crows are not unusual, and one raven in captivity survived to age twenty-nine. Anyway, corvids show no inclination toward breeding themselves up to huge numbers, filling the countryside with their kind (like the late passenger pigeon, or an infesting variety of insect) until conditions shift for the worse, and a vast population collapses. Instead, crows and their relatives reproduce at roughly the same stringent rate through periods of bounty or austerity, maintaining levels of population that are modest but consistent, and which can be supported throughout any foreseeable hard times. In this sense they are astute pessimists. One consequence of such modesty of demographic ambition is to leave them with excess time, and energy, not desperately required for survival.

The other thing they possess in excess is brain-power. They have the largest cerebral hemispheres, relative to body size, of

any avian family. On various intelligence tests—to measure learning facility, clock-reading skills, the ability to count—they have
made other birds look doltish. One British authority, Sylvia Bruce
Wilmore, pronounces them "quicker on the uptake" than certain
well-thought-of mammals like the cat and the monkey, and admits
that her own tamed crow so effectively dominated the other animals in her household that this bird "would even pick up the
spaniel's leash and lead him around the garden!" Wilmore also
adds cryptically: "Scientists at the University of Mississippi have
been successful in getting the cooperation of Crows." But she
fails to make clear whether that was as test subjects, or on a
consultative basis.

From other crow experts come the same sort of anecdote.
Crows hiding food in all manner of unlikely spots and relying on
their uncanny memories, like adepts at the game of Concentration,
to find the caches again later. Crows using twenty-three distinct
forms of call to communicate various sorts of information to each
other. Crows in flight dropping clams and walnuts on highway
pavement, to break open the shells so the meats can be eaten.
Then there's the one about the hooded crow, a species whose
range includes Finland: "In this land Hoodies show great initiative
during winter when men fish through holes in the ice. Fishermen
leave baited lines in the water to catch fish and on their return
they have found a Hoodie pulling in the line with its bill, and
walking away from the hole, then putting down the line and
walking back on it to stop it sliding, and pulling it again until
[the crow] catches the fish on the end of the line." These birds
are bright.

And probably—according to my theory—they are too bright
for their own good. You know the pattern. Time on their hands.
Under-employed and over-qualified. Large amounts of potential
just lying fallow. Peck up a little corn, knock back a few grass-
hoppers, carry a beak-full of dead rabbit home for the kids, then
fly over to sit on a fence rail with eight or ten cronies and watch
some poor farmer sweat like a sow at the wheel of his tractor. An
easy enough life, but is this *it?* Is this *all?*

If you don't believe me just take my word for it: Crows are bored.

And so there arise, as recorded in the case file, these certain . . . no, *symptoms* is too strong. Call them, rather, *patterns of gratuitous behavior*.

For example, they play a lot.

Animal play is a reasonably common phenomenon, at least among certain mammals, especially in the young of those species. Play activities—by definition—are any that serve no immediate biological function, and which therefore do not directly improve the animal's prospects for survival and reproduction. The corvids, according to expert testimony, are irrepressibly playful. In fact, they show the most complex play known in birds. Ravens play toss with themselves in the air, dropping and catching again a small twig. They lie on their backs and juggle objects (in one recorded case, a rubber ball) between beak and feet. They jostle each other sociably in a version of "king of the mountain" with no real territorial stakes. Crows are equally frivolous. They play a brand of rugby, wherein one crow picks up a white pebble or a bit of shell and flies from tree to tree, taking a friendly bashing from its buddies until it drops the token. And they have a comedy-acrobatic routine: allowing themselves to tip backward dizzily from a wire perch, holding a loose grip so as to hang upside down, spreading out both wings, then daringly letting go with one foot; finally, switching feet to let go with the other. Such shameless hot-dogging is usually performed for a small audience of other crows.

There is also an element of the practical jokester. Of the Indian house crow, Wilmore says: ". . . this Crow has a sense of humor, and revels in the discomfort caused by its playful tweaking at the tails of other birds, and at the ears of sleeping cows and dogs; it also pecks the toes of flying foxes as they hang sleeping in their roosts." This crow is a laff riot. Another of Wilmore's favorite species amuses itself, she says, by "dropping down on sleeping rabbits and rapping them over the skull or settling on drowsy cattle and startling them." What we have here is actually a distinct subcategory of playfulness known, where I come from at least, as Cruisin' For A Bruisin'. It has been clinically linked to boredom.

Further evidence: Crows are known to indulge in sunbathing. "When sunning at fairly high intensity," says another British corvidist, "the bird usually positions itself sideways on to the sun and erects its feathers, especially those on head, belly, flanks and rump." So the truth is out: Under those sleek ebony feathers, they are tan. And of course sunbathing (like ice-fishing, come to think of it) constitutes prima facie proof of a state of paralytic ennui.

But the final and most conclusive bit of data comes from a monograph by K. E. L. Simmons published in the *Journal of Zoology*, out of London. (Perhaps it's for deep reasons of national character that the British lead the world in the study of crows; in England, boredom has great cachet.) Simmons's paper is curiously entitled "Anting and the Problem of Self-Stimulation." *Anting* as used here is simply the verb (or to be more precise, participial) form of the insect. In ornithological parlance, it means that a bird—for reasons that remain mysterious—has taken to rubbing itself with mouthfuls of squashed ants. Simmons writes: "True anting consists of highly stereotyped movements whereby the birds apply ants to their feathers or expose their plumage to the ants." Besides direct application, done with the beak, there is also a variant called *passive anting:* The bird intentionally squats on a disturbed ant-hill, allowing (inviting) hundreds of ants to swarm over its body.

Altogether strange behavior, and especially notorious for it are the corvids. Crows avidly rub their bodies with squashed ants. They wallow amid busy ant colonies and let themselves become acrawl. They revel in formication.

Why? One theory is that the formic acid produced (as a defense chemical) by some ants is useful for conditioning feathers and ridding the birds of external parasites. But Simmons cites several other researchers who have independently reached a different conclusion. One of these scientists declared that the purpose of anting "is the stimulation and soothing of the body," and that the general effect "is similar to that gained by humanity from the use of external stimulants, soothing ointments, counter-irritants (including formic acid) and perhaps also smoking." Another compared anting to "the human habits of smoking and drug-taking" and maintained that "it has no biological purpose but is indulged in for its own sake, for the feeling of well-being and ecstasy it induces . . . ."

You know the pattern. High intelligence, large promise. Early success without great effort. Then a certain loss of purposefulness. Manifestations of detachment and cruel humor. Boredom. Finally the dangerous spiral into drug abuse.

But maybe it's not too late for the corvids. Keep that in mind next time you run into a raven, or a magpie, or a crow. Look the bird in the eye. Consider its frustrations. Try to say something stimulating.

# VOX POPULI

*Frogs and Gorillas
in the Heart of the Heart of the Country*

News flash from the realm of sociometric erudition: A recent study conducted by two researchers from the School of Forestry and Environmental Studies at Yale University, under contract to the U.S. Fish and Wildlife Service, at indeterminate (but no doubt substantial) cost to us taxpayers, has authoritatively established that man's best friend is the dog.

Science on the march. The same study, directed by Dr. Stephen R. Kellert and published under the title, "Knowledge, Affection, and Basic Attitudes Toward Animals in American Society," also turned up a few other invaluable facts. For instance, we now have it on solid sociological evidence that one in every four Americans believes the manatee is an insect; that one person in two can't name the number of legs on a spider; that a third of the population believe snakes wear a thin covering of slime; and that frogs are more popular in our culture (though only barely) than gorillas. Such information, according to a Fish and Wildlife Service spokesman, "has important implications for wildlife conservation and management programs." As you can well imagine.

Dr. Kellert's legmen personally interviewed 3,107 adult Americans, each of whom had been randomly selected, soliciting from each a body of information that could be sorted into three general categories: (1) knowledge (or lack of) about animals; (2) attitudes toward animals; and (3) something vague and ineffable called "species preference." The first category was covered by questions like "True or false: Raptors are small rodents" and "True or false: Veal comes from lamb" and again "True or false: When frightened, an ostrich will bury its head in the sand." The results show that Americans these days are a little weak on zoological knowledge, but they make up for that with an abundance of attitudes. These were measured by statements—with which each person was invited to agree or disagree in various gradations of vehemence—such as "I have owned pets that were as dear to me as another person" and "I think a person sometimes has to beat a horse or dog to get it to obey orders properly." All this is lively enough stuff, sure, but the most intriguing part of the Kellert study is that pile of data assembled in Table 17, a single eloquent and suggestive page summarizing the findings on "species preference." "Wildlife management can never be a popularity contest," the director of the Fish and Wildlife Service has said unconvincingly, in a press release ballyhooing the Kellert report, one portion of which is designed to serve precisely that purpose.

Table 17 is nothing less (if Dr. Kellert has done his work well) than our national popularity roster of wildlife. Thirty-three different animals are therein rated against each other according to the relative testimonials of affection, or aversion, they received from those 3,107 putatively average American humans. At the very top of this invidious listing, I am bored to report, is the common domestic dog. Fido: object of our most ardent interspecific devotion. I suppose it's only natural in a country where Rin-Tin-Tin, Lassie and Huckleberry Hound have all, like Ted Koppel, had their own TV shows. Ranked in successive order after the dog come the horse (also a proven star of screen and tube), the swan (the swan? a majority of Americans have probably never seen one in living flesh), and the robin (a seedy unexceptional thrush known by the scientific name *Turdus migratorius*, which is best left untranslated). Ranking fifth by no surprise is the butterfly, most vapidly prettified of all insects. But take heart. After the appalling banality of taste reflected in those leading five, things begin looking up. Elected to sixth place in our national popularity contest was the trout.

Now I confess that this choice leaves me with some misgiving. Can it be that Dr. Kellert's sample of 3,000 dog-lovers might truly appreciate the grace, the subtle beauty, the courage and wit, the exemplary nobility of a wild river-dwelling trout? Or were they just voting their palates? Alack, I suspect that if the Maine lobster or the cherrystone clam had been nominated among this group of wildlife, either of them would have edged out the trout. In seventh position comes the salmon, another magisterial beast rather highly favored, but Dr. Kellert doesn't specify whether that's canned salmon or smoked.

Arguably, in a hierarchy of this sort, the only real honor lies near the bottom. Down there the rankings are more spirited if no less predictable. For instance the seventh-from-the-last animal on the chart, seventh-from-the-least-popular critter in the minds of our countrymen, is the vulture. Here we have a fact worth knowing. After the buzzard, in further-decreasing favor, appear the bat, the rattlesnake, the wasp, the rat, the mosquito and finally, dead last, the cockroach. It sounds like a rogues' pantheon for verminous creatures of especially distinguished character—or a list of chapter subjects for this book.

Far more puzzling, and open to speculative interpretation, is the middle region of Kellert's roster. What does it *mean*, that the gorilla (in position twenty) is only a shade less "preferred," by 3,000 representative Americans, than the (nineteen) frog? What is signified by the fact that people feel marginally greater goodwill toward the moose (number fifteen) than toward the poor harried whale (sixteen)? On what basis did those sampled folk resoundingly favor elephants (nine) over walruses (seventeen)? Most perplexing of all, how did cats end up down there in *twelfth place?*

Yes, twelfth. That's the Kellert report's biggest mystery: dogs rated first, cats rated an ignominious twelfth. Which credits them with just slightly less popularity, in the hearts of America, than (number eleven) turtles. This man from Yale tells us that American society venerates cats not quite so highly as it venerates turtles.

Obviously he is insane.

Obviously it can't be so. Anyone who cohabits with felines, anyone who has ever been owned by a cat, knows it can't be so. Anyone who runs a supermarket, or a rug-shampoo business, or a drapery-repair operation, or a mobile vet clinic offering house calls for distemper vaccination (like the one I wrote a check to last week), knows it can't be so. Certainly anyone who happens to glance on occasion at the *New York Times* best-seller list knows it can't be so—because *that* list has faithfully chronicled how cat-lovers across the country, hopeless and doting and demented cat-lovers, turned a man named Jim Davis into a billionaire and kept hundreds of bookstores afloat for months at a time. Back at the end of 1982, in fact, the *Times*'s collation of all sales figures for the year showed that fully *five* of the top fifteen bookstore paperbacks were titles devoted to Davis's cutesy cartoon cat. That amounts to more book purchases than Robert Ludlum, Jane Fonda and Richard Simmons *together* accounted for during the same year. Meanwhile, the only canine book to place at all in the *Times* figures was something called *No Bad Dogs: The Woodhouse Way*, evidently a treatise on taxidermy. Twelfth place, indeed.

So there simply must be a glitch in Dr. Kellert's statistical model. There must be a spitball masquerading as a distributional curve. And I think I know what it is.

Cat-haters. Feliphobes. Of these I suspect there are almost as many as there are cat-lovers; almost as much total megawattage of passion expended in despising the feline species as in simpering over it. Cats seem for some reason to inspire that polarization. Most people are strongly for or strongly against, with no middle ground. Dogs are more innocuous and less controversial: Aside from the malice with which any healthy person views Dobermans and black toy poodles, they evoke nothing worse than indifference. Not so with cats. Cats inspire venom, determined callousness, too often even shocking brutality. It may be a function of that notorious feline hauteur (cats being capable of every emotion we might expect from a pet except loyalty and guilt). It may be a matter of allergies. It may be something much more elusive and atavistic that goes back beyond pharaonic Egypt. Or it may be jealousy over the paperback royalty figures. Anyway, and whatever the cause, many people simply hate cats.

This hatred could have been measured into the balance—along with the ill-defined entity labeled "preference"—by Dr. Kellert's rating methodology: It was possible to vote strongly against a particular species, as well as strongly for it. And the standard deviation (a statistical measure of variance) in Kellert's cat vote shows that that is precisely what happened. Furthermore, cat-haters in the sample group were no doubt pushing the cat still lower, into twelfth place, by awarding insincere crossover votes to those swans and elephants and turtles. That's my theory, at least, about Dr. Kellert's data. Cats weren't bested in a fair popularity race. They were blackballed. It shouldn't happen to a dog.

Still, unanswered questions remain. Other uncertainties, raised by the Kellert study and having "important implications for wildlife conservation," linger on. So in closing I'll try to clear them up here.

First, the manatee is in fact a tiny bird that merely *resembles* an insect. A spider generally has seven legs, or (if you count the eyes) fifteen. The skin of a snake is covered, not with slime, but with a coating of finely powdered graphite. Rabbits, not raptors, are small rodents. Veal comes from Kansas City. When frightened, a Kuwaiti, not an ostrich, will bury his head in the sand.

Finally, the matter of why frogs should be more popular than gorillas, in American culture, is still without adequate explanation. But sociology is at work on the problem.

# RUMORS
# OF A SNAKE

---

*Pondering
the Anaconda at Great Length*

What this world needs is a good vicious sixty-foot-long Amazon snake.

Don't look at me, it's not my private notion. There is a broader mandate of some murky sort, not biological but psychic, issuing from that delicately balanced ecosystem we call the human mind. *Give us a huge snake. A monster, a serpent out at the far fringe of imaginability. Let it inhabit the funkiest jungle. A horrific thing, slithering along in elegant menace, belly distended with pigs and missing children.* The evidence of this odd yearning is oblique but cogent: Lacking any such beast, we are eager to settle for rumors of one. Otherwise how to explain the breathless compoundment of hearsay, tall tale, and exaggeration that has always surrounded the anaconda?

A fair example appears in the memoirs of Major Percy Fawcett. In 1906 he was sent out from London by the Royal Geographical Society to make a survey along certain rivers in western Brazil. "The manager at Yorongas told me he killed an anaconda 58 feet long in the Lower Amazon. I was inclined to look on this as an exaggeration at the time, but later, as I shall tell, we shot one even larger than that." The disclaimer is a cagey stroke. Major Fawcett would have us take him for a hard-headed British skeptic.

Later he tells: "We were drifting easily along in the sluggish current not far below the confluence of the Rio Negro when almost under the bow of the *igarité* there appeared a triangular head and several feet of undulating body. It was a giant anaconda. I sprang for my rifle as the creature began to make its way up the bank, and hardly waiting to aim smashed a .44 soft-nosed bullet into its spine, ten feet below the wicked head. . . . We stepped ashore and approached the reptile with caution. It was out of action, but shivers ran up and down the body like puffs of wind on a mountain tarn. As far as it was possible to measure, a length of forty-five feet lay out of the water, and seventeen feet in it, making a total length of sixty-two feet." The indisputable logic of good arithmetic. It might all be true but most likely it isn't.

An adventurer of the 1920s named F.W. Up de Graff offered a similar account, having observed his anaconda in shallow water. "It measured fifty feet for certainty, and probably nearer sixty. This I know from the position in which it lay. Our canoe was a twenty-four footer; the snake's head was ten or twelve feet beyond the bow; its tail was a good four feet beyond the stern; the center of its body was looped up into a huge S, whose length was the length of our dugout and whose breadth was a good five feet." But size estimates made in a watery medium are notoriously unreliable—especially when that watery medium is the Amazon River.

Bernard Heuvelmans, in his feverish book about stalking mysterious animals, tells at third or fourth hand of another Brazilian specimen that was purportedly killed in 1948. "The snake, which was said to measure 115 feet in length, crawled ashore and hid in the old fortifications of Fort Tabatinga on the River Oiapoc in the Guaporé territory. It needed 500 machine-gun bullets to put paid to it. The speed with which bodies decompose in the tropics and the fact that its skin was of no commercial value may explain why it was pushed back in the stream at once." Always for the gigantic individuals there is this absence of physical evidence, and always a waterproof reason for the absence: no camera on hand, rotting meat, even the skin was too heavy to carry out. One photograph did exist that, during the 1950s, was sold all over Brazil as a postcard, its caption claiming a length of 131 feet for the snake pictured. Unfortunately, no object of reference appeared in the photo with it. That snake might as easily have been a robust but minuscule twenty-footer.

My own modest sighting comes not from Brazil but from northeastern Ecuador, along the Rio Aguarico, in a remote zone of lowland jungle that may be as favorable to the production and growth of anacondas as almost anywhere in the Amazon drainage. Like those other wide-eyed witnesses Fawcett and Up de Graff, I was in a dugout canoe. Our guide was an intrepid and jungle-smart young man named Randy Borman, who spotted the big snake on a log tangle near the river bank while the rest of us were gawking elsewhere. He steered the boat in for a closer look.

Dark gunmetal gray with sides mottled in reddish brown, the anaconda was sunning itself placidly. Barely above the water on a low-riding log; protectively colored and patterned so that, even from ten yards away, it was virtually invisible. Randy edged the canoe closer. Do you see it now? Some of us did, several admitted they didn't. We moved closer. Here indeed was a formidable snake. Still motionless, still sunbathing dreamily. I was delighted with this glimpse of an anaconda in the wild but—amid the soft brushfire crackle of camera shutters—we had already ventured closer than I ever expected to get. Then closer still. A very tolerant and self-possessed snake. Beautiful big head. Thick graceful coils of body. Sizable brown eye. Hello, Randy? Just when I thought our guide would back the canoe off, instead he dove over the gunnel to grab this creature around the neck.

A deeply startling act. But Randy came up again, deftly, with a great armload of anaconda wrapping itself onto him in surprise and anger, squeezing with the authority of a species that does its killing by constriction. My face bore the contemplative expression of two eggs sunnyside in a white Teflon skillet.

Randy smiled calmly. "We'll take him back to camp for the others to see." An hour later, having been much fawned over and photographed, the animal was gently released back into its river. A few fast pulses of undulant swimming, then a dive beneath the brown water, and it was gone. There was a convenient absence of physical evidence.

So in recounting the story afterward (which I have not hesitated to do often, cornering people at parties and hoping the talk might turn to giant reptiles) I could make that poor snake any damn size I pleased. A piddling ten feet? Maybe eleven? Roughly the same girth as a man's biceps? In fact (so I would say, with engaging dismissiveness) it was rather dainty as this species goes. Still, an impressive beast.

Very little is known about the biology of *Eunectes murinus*, the anaconda; even less about its life history in the wild; and a sad fact is that no one seems much to care. Not a single field-research project (one expert has told me) is currently being done on it. The species has not yet found its George Schaller, its Dian Fossey, its Jane van Lawick-Goodall.

We know that it is a nonvenomous constrictor of the boa family. We know that it is aquatic, preferring slow rivers and swamps. That it bears live young (as opposed to laying eggs) in litters of up to eighty. That it is native to tropical South America east of the Andes, and also to the island of Trinidad. That (unlike the other boas and pythons) it does poorly and dies soon in captivity. But as to the rest, its favored diet, its daily and seasonal rhythms, its mating and birthing behavior, physiology, growth rate, optimal longevity: almost a total blank. There is an absence of evidence.

Admittedly, the prospect of studying full-grown anacondas in their own habitat offers an array of uniquely forbidding logistical problems. For that reason or whatever others, scientific consideration of *Eunectes murinus* has been limited almost entirely to the same simple question that so mesmerized those early explorers: *How big does it get?* Well, really quite big. Bigger than any other snake on earth. *But how big is that?*

A second sad fact about the anaconda: By scientific standards of verification, it just doesn't seem to be nearly so large as everyone seems to want to believe it is. Forget 131 feet. Forget 62 feet, even with faultless arithmetic. Discount the record-length skins, which generally have been stretched by a good 20 percent in the process of tanning. Scientists have their own unstretchable views on this matter.

One respected herpetologist, Afrânio do Amaral, has posited a maximum length for the anaconda of about forty-two feet. But then Afrânio do Amaral is a Brazilian, arguably with a vested patriotic interest. And after him the figures only get stingier. James A. Oliver of the American Museum of Natural History was willing to grant 37½ feet, based on the measurement made by a petroleum geologist, with a surveyor's tape, of a snake shot along the Orinoco River. But again in this case there was the problem of physical evidence: "When they returned to skin it, the reptile was gone," we are told. "Evidently it had recovered enough to crawl away." Teddy Roosevelt is said to have offered $5,000 for a skin or skeleton thirty feet long, and the money was never claimed. Sherman and Madge Minton, authors of several reliable snake books, declare that "To the best of our knowledge, no anaconda over twenty-five feet long has ever reached a zoo or museum in the United States or Europe." And Raymond Ditmars, an eminent snake man at the Bronx Zoo early in this century, wouldn't believe anything over nineteen feet.

Can these people all be discussing the same animal? Can Ditmars's parsimonious nineteen feet be reconciled with the eyewitness account of Major Fawcett? Does Roosevelt's unclaimed cash square with Heuvelmans's 115 feet of worthless rotting meat? It seems impossible.

But new evidence has lately reached me that suggests an explanation for everything. The evidence is a small color photograph. The explanation is relativity.

Not Einstein's variety, but a similar sort, which I shall call *Amazonian relativity*. It's very simple: The *true genuine size* of an anaconda (this theory applies equally well to piranha and bird-eating spiders) is relative to three other factors: (1) whether or not the snake is alive; (2) how close you yourself are to it; and (3) how close both of you are, at that particular moment, to the Amazon heartland. A live snake is always bigger than a dead one, even allowing for posthumous stretch. And as the other two distances decrease—from you to the snake, from you to the Amazon—the snake varies inversely toward humongousness.

This small color photograph, of such crucial scientific significance, arrived in the mail from an affable Dutch-born engineer, a good fellow I met on that Rio Aguarico trip. Unlike me, he carried a camera; sending the print was meant as a favor. In its foreground can be seen the outline of my own dopey duck-billed hat. The background is a solid wall of green jungle. At the center of focus is Randy Borman, astride the stern of his dugout, holding an anaconda. Dark gunmetal gray with sides mottled in reddish brown.

The snake is almost as big around as his wrist. It might be five feet long. Possibly close to six. But photographs can be faked. I don't believe this one for a minute.

# AVATARS OF
# THE SOUL IN
# MALAYA

---

*Moth and Butterfly,*
*Fact and Idea*

Consider now the Lepidoptera, in all their vacant splendor.

They are the bimbos of the natural world: more beautiful and less interesting, arguably, than any other order of animals. An evolutionary experiment in sheer decorative excess, with a high ratio of surface to innards. They move through the air like pulses of idle thought. They have a weakness for flowers. They are prodigiously diverse without being adventurous: roughly 150,000 known species, all of which behave pretty much alike; 150,000 distinct patterns, but in each case a six-legged worm strung between kites. They are silent. Detached and diaphanous. Generally they possess neither teeth nor jaws. They feed pacifically on plant liquids or (some species) just go hungry through their entire adulthood. Fly on wings that are fleshless and papery, flashing bright iridescent colors produced by the devious exploitation of tiny prisms and mirrors. Certainly these are real physical creatures, yes; then again they just don't seem to be quite all *there*. Aristotle was on to something, I think, when in the fourth century B.C.

47

he used the Greek-alphabet equivalent of the word "psyche" to mean both *soul* and *butterfly*.

They might be insects. Or they might be Platonic ideas.

In classical Greece, and then later in Rome, this link with the spiritual realm was applied to both groups within the Lepidoptera, moths as well as butterflies. Both moths and butterflies were delicate enough to suggest a pure being, freed of the carnal envelope. Both were known to perform a magical metamorphosis— from fat ugly caterpillar to gorgeous airborne adult, with a dormant pupal stage in between—that put humans in mind, especially, of resurrection from the grave. Moths may have been even more suited than butterflies to bearing this burden of symbolism, in that moths fly at night, like the souls of the deceased. Tomb-sculpture designs from imperial Rome have survived (thanks to later Italian scholars who copied them before the original stones were lost), on which appear butterflies and moths carved to represent the departing souls of the dead. And the motif has endured. One marker from a nineteenth-century grave in Massachusetts, for instance, shows a common monarch, *Danaus plexippus*, freshly emerged from its chrysalis and winging away. The soul as butterfly.

Evidently mankind has taken some small comfort, over the past couple millennia, from gazing upon Lepidoptera and positing this odd connection between their substance and our essence. Why the Lepidoptera? Because they are detached and diaphanous. Because their beauty is of an otherworldly sort. No sting, no bite, no bothersome buzz. Strict vegetarians. They represent an ideal of sweetness and gentle grace that seems almost innocent of the whole ruthless Darwinian free-for-all. No wonder they are, zoologically, so godawful boring to contemplate.

But cheer up. That's only the traditional, happytime view of the Lepidoptera. It applies to no more than about 149,990 species. And it takes no account whatsoever of a small Malayan jungle moth called *Calpe eustrigata*.

Here finally is a moth with character, a moth with *involvement*, a moth unafraid to get its knees dirty. The only one of its kind known in the world. *Calpe eustrigata* sucks blood from humans.

*   *   *

There is no common name for *C. eustrigata*, but it belongs to a large family of drab little moths known as the noctuids, notable in this country mainly for the damage that larvae of some species do to vegetables and grain. Your basic cutworms and earworms and celery-loopers. The more dreary noctuids are plentiful, but *C. eustrigata* itself is quite rare. It was discovered—almost accidentally—by an admirable fanatic named Hans Bänziger, a Swiss entomologist who was spending two years in the jungles of Thailand for research on a different group of noctuids.

In Thailand the work progressed satisfactorily; the moths behaved as they were expected to. No sign of any such creature as *C. eustrigata*. Then, toward the end of the two years, Bänziger got down into Malaya, where he wanted to investigate several species that were opportunistic blood-drinkers, of a purely nonaggressive sort. These Malayan moths were known to lick at the open wounds of large mammals, and to follow after mosquitoes (which are greedy and slovenly in the way they extract blood), lapping up what was spilled. Again the moths behaved as expected. At least they did until, late one night, Bänziger captured a particular specimen. He found it alit on a water buffalo.

Bänziger's own account, from a back issue of the journal *Fauna*: "I had become suspicious of this insect species because of a photograph taken a few days before while it was feeding on a Malay tapir. The photograph showed something very strange about the moth's proboscis. Now with a live specimen I intended to study its feeding behavior on myself. That night was to become especially exciting! Having incised my finger with a scalpel to draw fresh blood, I offered my finger to the caged moth. The moth climbed onto my finger and did in fact plunge its proboscis into the blood, but it appeared to imbibe none. Instead it stuck its straight, lancelike proboscis into the wound and, without any regard for the donor, penetrated the flesh. The pain I felt caused me to utter a cry of—joy!" Lucky the man who so loves his work. "I had discovered a moth which *pierces* to obtain blood." If you were a lepidopterist, you'd see that it was a pretty big moment in history.

During a month in Malaya, Bänziger found only twenty-four more specimens of *C. eustrigata*, but he kept them all busy poking and sucking at his hand. As an experimentalist, he lacked nothing by way of commitment. "Blood-sucking occupied from 10 to 60 minutes and blood continued to flow out of the wound for a few minutes after its cessation. Hours and sometimes even days later, the wound was still itching." The remarkable aspects of all this involved not just the matter of *behavior* (including Bänziger's behavior) but the matter of *anatomy*. Lepidopteran mouth structures are totally different from those of, say, a mosquito. The standard equipment for Lepidoptera is a long flexible tube that remains rolled up in a coil under the head until, when needed, it can be sprung out straight by hydraulic pressure, like one of those paper squeakers in the mouth of a drunk on New Year's Eve. In extended position it allows the insect to suck nectar from the reservoir of a deep flower. But this thing is a drinking straw, not a drill. Until Bänziger, no lepidopterist had ever seen a moth whose proboscis could be stabbed through the human skin.

There was also the mystery of its phylogeny: Where, in an evolutionary sense, had *C. eustrigata* come from? How had the blood-sucking equipment evolved? What were the intermediate stages between Bänziger's new species and those other Lepidoptera—all 149,990 of them—who noodle from blossom to blossom drinking nectar through their long delicate schnosters? What manner of temptation could have lured certain moth species astray, turning their taste from flowers to blood?

The answer, evidently, was fruit. Faced with mortal competition over limited supplies of nectar, a number of noctuid species have adapted themselves to feeding upon the juice of overripe fruit. Some have developed stronger and sharper mouth tubes that allow them to pierce the skin of soft fruits like peaches and raspberries, and suck out their fill of juice. One species is even armed with a proboscis that will penetrate the skin of an orange. Among these fruit-piercing moths are several close cousins to *C. eustrigata*. Bänziger suggests that, in a habitat where fruit was available only seasonally but juicy mammals were present year round, desperate necessity might have led to the next logical step: vampirism, as practiced by *C. eustrigata*.

But the vampire moth was just a distraction from what had brought Hans Bänziger out to Southeast Asia. He was there to study a group of species he called, rather blandly, the "eye-frequenting" Lepidoptera. Moths that live on a diet of tears.

Literally. These wondrous creatures may have evolved from the opportunistic blood-drinkers that clean up after messy mosquitoes. Bänziger says: "Probably by crawling about on their mammalian hosts some moths found the eyes, where there are always discharges. And thus there evolved the habit of dining exclusively upon eye discharges, which contain various proteins such as globulin, albumen, and others in the leucocytes and epithelial cells in tears." These moths would drink from the eyes of elephants. They would drink from the eyes of horses, buffalo, antelope, deer, pigs. And as Bänziger demonstrated, with his characteristic élan, they would drink from the eyes of man. "The lachrymal secretion was very much stimulated by the activity of the moth. After 30 min. my eye was so irritated that I was forced to interrupt the experiment." But not before the specimen had drunk freely from Bänziger's own tears of—joy! There is even a photograph, showing a small noctuid of the species *Lobocraspis griseifusa* perched head-down across Bänziger's brow, its tube extended to drink fluid off the surface of his cornea. With a caption: "Note the deep penetration of the proboscis between eye and eye lid."

Note the deep penetration of Lepidoptera between fact and imagination. I suspect the Greeks and the Romans would have known what to make of *Lobocraspis griseifusa*, a species in the spiritual tradition. Imagine how Ovid would have loved them. The souls of the dead return, on powdered wings and in silence, to comfort mankind. To condole with us, who remain behind. To drink away our very tears.

It's a nice thought; too nice to be true. And besides, what would we do then about *C. eustrigata?*

# A REPUBLIC
# OF COCKROACHES

*When the Ultimate Exterminator*
*Meets the Ultimate Pest*

In the fifth chapter of Matthew's gospel, Christ is quoted as saying that the meek shall inherit the earth, but other opinion lately suggests that, no, more likely it will go to the cockroaches.

A decidedly ugly and disheartening prospect: our entire dear planet—after the final close of all human business—ravaged and overrun by great multitudes of cockroaches, whole plagues of them, whole scuttering herds shoulder to shoulder like the old herds of bison, vast cockroach legions sweeping as inexorably as driver ants over the empty prairies. Unfortunately this vision is not just the worst Kafkaesque fantasy of some fevered pessimist. There is also a touch of hard science involved.

The cockroach, as it happens, is a popular test subject for laboratory research. It adapts well to captivity, lives relatively long, reproduces quickly, and will subsist in full vigor on Purina Dog Chow. The largest American species, up to two inches in length and known as *Periplaneta americana*, is even big enough for easy dissection. One eminent physiologist has written fondly: "The laboratory investigator who keeps up a battle to rid his rat colony of cockroaches may well consider giving up the rats and working with the cockroaches instead. From many points of view the roach is practically made to order as a laboratory subject. Here is an animal of frugal habits, tenacious of life, eager to live in the laboratory and very modest in its space requirements." Tenacious of life indeed. Not only in kitchen cupboards, not only among the dark corners of basements, is the average cockroach a hard beast to kill. Also in the laboratory. And so also it would be, evidently, in the ashes of civilization. Among the various biological studies for which cockroaches have served as the guinea pigs—on hormone activity, parasitism, development of resistance against insecticides, and numerous other topics—have been some rather suggestive experiments concerning cockroach survival and atomic radiation.

Survival. Over the centuries, over the millennia, over the geologic epochs and periods and eras, that is precisely what this animal has proved itself to be good at. The cockroach is roughly 250 million years old, which makes it the oldest of living insects, possibly even the oldest known air-breathing animal. Admittedly "250 million years" is just one of those stupefying and inexpressive paleontological numbers, so think of it this way: Long before the first primitive mammal appeared on earth, before the first bird, before the first pine tree, before even the reptiles began to assert themselves, cockroaches were running wild. They were thriving in the great humid tropical forests that covered much of the Earth then, during what geologists now call the Carboniferous period (because so much of that thick swampy vegetation was eventually turned into coal). Cockroaches were by far the dominant insect of the Carboniferous, outnumbering all other species together, and among the most dominant of animals. In fact, sometimes this period is loosely referred to as the Age of Cockroaches. But unlike the earlier trilobites, unlike the later dinosaurs, cockroaches lingered on quite successfully (though less obtrusively) long after their heyday—because, unlike the trilobites and the dinosaurs, cockroaches were versatile.

They were generalists. Those primitive early cockroaches possessed a simple and very practical anatomical design that remains almost unchanged in the cockroaches of today. Throughout their evolutionary history they have avoided wild morphological experiments like those of their near relatives, the mantids and walking sticks, and so many other bizarrely evolved insects. For cockroaches the byword has been: Keep it simple. Consequently today, as always, they can live almost anywhere and eat almost anything.

Unlike most insects, they have mouthparts that enable them to take hard foods, soft foods, and liquids. They will feed on virtually any organic substance. One study, written a century ago and still considered authoritative, lists their food preferences as "Bark, leaves, the pith of living cycads [fern palms], paper, woollen clothes, sugar, cheese, bread, blacking, oil, lemons, ink, flesh, fish, leather, the dead bodies of other Cockroaches, their own cast skins and empty egg-capsules," adding that "Cucumber, too, they will eat, though it disagrees with them horribly." So much for cucumber.

They are flattened enough to squeeze into the narrowest hiding place, either in human habitations or in the wild. They are quick on their feet, and can fly when they need to. But the real reason for their long-continued success and their excellent prospects for the future is that, beyond these few simple tools for living, they have never specialized.

It happens to be the very same thing that, until recently, could be said of *Homo sapiens*.

Now one further quote from the experts, in summary, and because it has for our purposes here a particular odd resonance. "Cockroaches," say two researchers who worked under sponsorship of the United States Army, "are tough, resilient insects with amazing endurance and the ability to recover rapidly from almost complete extermination."

It was Jonathan Schell's best-selling jeremiad *The Fate of the Earth*, published in 1982, that started me thinking about cockroach survival. *The Fate of the Earth* is a very strange sort of book, deeply unappealing, not very well written, windy and repetitious, yet powerful and valuable beyond measure. In fact, it may be the dreariest piece of writing that I ever wished everyone in America would read. Its subject is, of course, the abiding danger of nuclear Armageddon. Specifically, it describes in relentless scientific detail the likelihood of total human extinction following a full-scale nuclear war. In a section that Schell titles "A Republic of Insects and Grass," there is a discussion of the relative prospects for different animal species surviving to propagate again after mankind's final war. Schell takes his facts from a 1970 symposium held at Brookhaven National Laboratory, and in summarizing that government-sponsored research he says:

> For example, the lethal doses of gamma radiation for animals in pasture, where fallout would be descending on them directly and they would be eating fallout that had fallen on the grass, and would thus suffer from doses of beta radiation as well, would be one hundred and eighty rads [a standard unit of absorbed radiation] for cattle; two hundred and forty rads for sheep; five hundred and fifty rads for swine; three hundred and fifty rads for horses; and eight hundred rads for poultry. In a ten-thousand-megaton attack, which would create levels of radiation around the country averaging more than ten thousand rads, most of the mammals of the United States would be killed off. The lethal doses for birds are in roughly the same range as those for mammals, and birds, too, would be killed off. Fish are killed at doses of between one thousand one hundred rads and about five thousand six hundred rads, but their fate is less predictable. On the one hand, water is a shield from radiation, and would afford some protection; on the other hand, fallout might concentrate in bodies of water as it ran off from the land. (Because radiation causes no pain, animals, wandering at will through the environment, would not avoid it.) The one class of animals containing a number of species quite likely to survive, at least in the short run, is the insect class, for which in most known cases the lethal doses lie between about two thousand rads and about a hundred thousand rads. Insects, therefore, would

be destroyed selectively. Unfortunately for the rest of the environment, many of the phytophagous species [the plant-eaters] . . . have very high tolerances, and so could be expected to survive disproportionately, and then to multiply greatly in the aftermath of an attack.

Among the most ravaging of those phytophagous species referred to by Schell is an order of insects called the Orthoptera. The order Orthoptera includes locusts, like those Moses brought down on Egypt in plagues. It also includes crickets, mantids, walking sticks, and cockroaches.

Ten thousand rads, according to Schell's premises, is roughly the average dosage that might be received by most living things during the week immediately following Armageddon. By coincidence, 10,000 rads is also the dosage administered to certain test animals in a study conducted, some twenty-four years ago, by two researchers named Wharton and Wharton. The write-up can be found in a 1959 volume of the journal *Radiation Research*. The experiment was performed under the auspices, again, of the U.S. Army. The radiation was administered from a two-million-electron-volt Van de Graaff accelerator. The test animals were *Periplaneta americana*, those big American cockroaches.

Remember now, a dose of 180 rads is enough to kill a Hereford. A horse will die after taking 350 rads. The average lethal dose for humans isn't precisely known (because no one is performing quite such systematic experiments on humans, though again the Army has come closest, with those hapless GIs forced to ogle detonations at the Nevada Test Site), but somewhere around 600 rads seems to be a near guess.

By contrast, cockroaches in the laboratory dosed with 830 rads routinely survive to die of old age. Their *average* lethal dose seems to be up around 3,200 rads. And of those that Wharton and Wharton blasted with 10,000 rads, *half* of the group were still alive two weeks later.

The Whartons in their *Radiation Research* paper don't say *how much* longer those hardiest cockroaches lasted. But it was long enough, evidently, for egg capsules to be delivered, and hatch, and for the cycle of cockroach survival and multiplication—unbroken throughout the past 250 million years—to continue on. Long enough to suggest that, if the worst happened, cockroaches in great and growing number would be around to dance on the grave of the human species.

With luck maybe it won't happen—that ultimately ugly event foreseen so vividly by Jonathan Schell. With luck, and with also a gale of informed and persistent outrage by citizenries more sensible than their leaders. But with less luck, less persistence, what I can't help but envision for our poor raw festering planet, in those days and years after the After, is, like once before, an Age of Cockroaches.

# PROPHETS
# AND
# PARIAHS

# THE
# EXCAVATION
# OF JACK HORNER

"I don't give a shit *what* killed the dinosaurs," says John R. Horner. Strange talk for an eminent paleontologist, but not out of character for this particular one. He is exaggerating his natural brusqueness only a little, in the interest of stressing a point. "They dominated the earth for 140 million years. Let's stop asking why they *failed* and try to figure out why they *succeeded* so well." From Horner's perspective, the entire Mesozoic era—during which the dinosaurs first appeared, flourished, diversified, rose to supremacy among all terrestrial creatures and then, somewhat abruptly, disappeared—is a Horatio Alger story, not a murder mystery.

Jack Horner's perspective is unconventional but authoritative. His recent fossil discoveries, and the surprising deductions toward which those fossils have led him, are being followed raptly by paleontologists all over the world. With his scruffy beard, longish hair, balding pate, he looks like a skinnier and younger version of the late Warren Oates. On location, let's say, for a good-humored film about raucous and disreputable prospectors. But in fact, Horner is one of a trinity of men—John Ostrom and Robert Bakker are the others—who during the past fifteen years have been drastically reshaping our understanding of dinosaurs.

Ostrom is a venerable professor at Yale. Bakker has lately gone from Harvard to Johns Hopkins. Meanwhile Jack Horner sits, wearing a plaid flannel shirt and beaten-down running shoes, in the basement of a small museum in a place called Bozeman, Montana.

Like Richard Leakey, with his study of early mankind in northern Kenya, Horner has stepped suddenly into the front rank of scientists in his field despite the near-total lack of academic credentials. He never bothered to finish college. Never went to grad school. Doesn't read German or Russian. Knows almost nothing about computers. Unlike Leakey, Horner did not even have the advantage of famous scientist parents; his family owned a gravel-and-concrete business in Shelby, Montana. Horner is simply a brilliant and dogged bone-hunter, a field man, a natural, with a keen brain for imagining the ecological particulars of an age 70 million years gone. He has a nose for fossils, and a head full of provocative ideas.

On a bare hillside not far from the Teton River in north-western Montana, Horner and his field associates have unearthed a nest, roughly six feet across, containing the bones of eleven baby dinosaurs. In the same vicinity they have also found other nests, several more babies, and the fossilized remnants of more than 300 dinosaur eggs. Throughout the whole history of fossil collection, dinosaur eggs and juveniles have remained breathtakingly rare; no other nest full of hatchlings has *ever* been found. Consequently, there has been a tantalizing absence of just that particular sort of evidence necessary to answer certain crucial questions—questions about dinosaurian breeding habits, patterns and rates of growth, behavior among others of their kind. Jack Horner now has that sort of evidence.

Based on his finds, Horner believes that at least one group of dinosaurs were sociable, relatively intelligent, warm-blooded, and solicitous toward their own infant offspring. It's a little like announcing, 500 years ago, that the Earth isn't flat after all.

Warm-bloodedness, nesting in colonies, and extended parental care are all generally non-reptilian attributes, associated rather with mammals and birds. Reptiles are cold-blooded. They don't (except in rare and disputable cases) tend their young. They don't show advanced social behavior. Reptiles as we know them just don't act in the manner that Horner's nest-field seems to indicate.

But maybe the dinosaurs were not nearly so reptilian as tradition, and eight generations of paleontologists, have decreed. Maybe, suggests Jack Horner, they were something utterly different.

In more senses than one, Horner grew up among dinosaurs.

The wild country of Montana has always been a bone-digger's mecca, partly because its hillsides and gulches have remained almost undisturbed by human development, more basically because this happened to be a place where great numbers of dinosaurs lived and died. Toward the end of the Cretaceous period, 70 million years ago, what are now the Midwest and the Great Plains were covered by a vast inland seaway, and central Montana was its western seacoast. Dinosaurs thrived in that swampy coastal zone and, when an individual died, sediments washing down from the newly burgeoning Rocky Mountains were liable to bury it. Finally, the seaway withdrew, the Cretaceous sediments were overlain with more recent strata; as subsequent epochs passed, erosion cut down through those strata, crustal pressures buckled and tilted the land, and in many places the Cretaceous deposits were re-exposed to daylight. The result is a rich hunting-ground for fossils, an enormous bone-yard dating from exactly that time at which the dinosaurs hit their peak.

Back in 1855, the first dinosaur fossils to be found and described in the western hemisphere were taken from beds along the Judith River, not far from Fort Benton, Montana. In 1902, the modern world's first glimpse of *Tyrannosaurus rex* came from a dig near Jordan, south of Fort Peck. Jack Horner spent his boyhood at large in this terrain. He found his own first dinosaur bone when he was eight. A systematic kid, he used white paint to label the fist-sized chunk as specimen "104-A" among a boy's box of fossils.

"Did you save that bone?"

"Yup," Horner says.

"Do you still have it?"

"Yup."

"What is it? What part of what sort of animal?"

"I don't have the slightest idea."

Horner struggled through Shelby's only high school, and it would be understatement to say that in the classroom he showed no promise of future scientific renown. Languages, for some reason, were especially a problem. "Took me two years to manage a D in Latin One." Nevertheless he went on to the university, down at Missoula, hoping to do a degree in geology. His father harbored a dream, on Jack's behalf, of the career of a mining engineer. Jack himself was still dreaming about fossils. More specifically: about dinosaur fossils.

"Dinosaurs are really neat animals," he says even now, shamelessly ingenuous in his enthusiasm. "I mean, dinosaurs are *really neat* animals." Often enough he discusses them in the present tense, hypothesizing details about certain species or families as would any wildlife biologist: "A baby hadrosaur has very little to protect it."

But his initial try at the university ended sourly. "I'm a product of the sixties," Horner says often, and with a glint of perverse pride—and justifiably, since there is no better way to qualify than by what befell him next. He flunked out of college in 1965. And was drafted immediately by the Marines.

"Everybody thought the Marines didn't draft. Remember? That's what I thought too. The Marines?!"

They sent him through something called "para-frog" training in Okinawa, where Horner was taught how to leap out of airplanes, over water, wearing a parachute on one part of his body and scuba tanks on another. Characteristically upbeat about personal matters, he counts himself lucky: He was never required to jump into the ocean during combat. Instead he jumped into jungle. Most of his thirteen months in Vietnam were spent on "force recon" duty. He would be dropped into the DMZ or some other feverish corner of Vietnamese jungle, with a small team or alone, carrying minimal firepower but a strong radio, and simply stay out there, discreetly, avoiding combat but reporting back south about whomever and whatever he saw.

During one of these solitary patrols, near Quang Tri just south of the DMZ, he encountered a pair of North Vietnamese students. They were taking instruction at a Buddhist temple. Walking in out of the jungle, Horner had seen the temple and was curious. He set his rifle down at the front door, because that seemed the courteous thing, and entered. Several Buddhist monks were there, with this pair of students; the monks were teaching them English and a smattering of biological sciences. The two students, Horner recalls, were the first people in all Vietnam—Asian or American— with whom he could talk. "*Really* talk. About more than hat size," says Horner. "Or what an M-16 could do to the human body. 'Yew ever see what a M-16 kin do to the human body?' That was always a favorite. So these two students, well, we just started talking. They were, literally, the most intelligent people I met in Vietnam." He told them a little about himself. Told them he was from Shelby, Montana. One of the North Vietnamese students said: "Is that close to Butte?"

The best Vietnam duty of all, to Horner's taste, was when he was left by himself to spend a week or two on "recon station," manning an unprotected little lookout post in the midst of some ungodly forward zone. "I liked being alone in the jungle," he explains. Surrounded by exotic animals and crazy wild vegetation. A course of nature study. Not so different, he claims, from being home in the outback of Montana. Then one day he called in an artillery barrage, but gave the wrong coordinates, and collected a leg full of shrapnel when the American cannons shelled him instead of the enemy.

After Vietnam, Horner went back to the University of Montana, floundering as hopelessly as before. He was still fascinated by geology and paleontology, but for a degree in those subjects he was required also to pass courses in math, liberal arts, and a couple of foreign languages. The language requirements in particular were daunting. "I was in a Russian class for three days before I figured out it was second term." At one juncture, Horner recalls, his grade-point average was so low it could only be rounded off to zero. Out again, in again, out again, yet during all these years of frustrating academic travail, Horner was still going back up each summer, or whenever possible, to the Cretaceous formations in central Montana. He was digging and collecting with a fanaticism derived from sheer enjoyment. His determination, his love for being outdoors on the Montana landscape, his gift for reading rock, his stamina for crawling around in coulees on bruised hands and knees for hours at a time with a finely focused attention—all these were making him a highly experienced field paleontologist, whatever the college records might say.

In 1973 he left the university altogether, and began driving a gravel truck. Stone is a leitmotif throughout Horner's life.

Not long thereafter he moved up to an 18-wheel tractor-trailer rig, hauling tanks full of liquid fertilizer all over the state. He was paid by the day but there was one incentive for making good time. "I always kept an eye open for Cretaceous rock. When I'd come to what seemed like a fossilly area, I'd just stop, unhook my trailer, and drive off across the badlands in that tractor. To look for dinosaur bones." Yet the truck driving, even on these terms, was never in Horner's mind more than an interim situation. During the same period, he was mailing job-query letters to every paleontological museum in the country.

In 1975 he was hired by Princeton University to work as a fossil preparator (the paleontological equivalent of a dental technician), cleaning and glueing specimens that were then to be studied by other people. Faculty scientists. Folk with Ph.D.s. Horner was abundantly over-qualified as a preparator, having done the same sort of chores in support of his own private studies for most of the past two decades. Nevertheless he stayed at Princeton for seven years, polishing his skills, learning the ways of museum work, earning a little autonomy, expanding his role by increments, and getting up and down the East Coast for a close look at every important dinosaur collection from Harvard to the

Smithsonian. He also spent his vacations each summer out in Montana, gathering more fossils from the gulches and bluffs he knew well, and thereby greatly enriching the Princeton collection.

One other significant matter was unearthed during those Princeton years, not a fossil but a fact. Thanks to a campus poster and then an exam, Horner learned for the first time that he suffered—and always had—from dyslexia. It cast some light on the inaptitude for languages, the academic struggles, the strong preference toward field work. Words on a page shifted and twisted and tangled themselves before Jack Horner's eyes. But a bone was a thing of solidity and eloquence.

In 1978, still under the Princeton aegis, he went back to northwestern Montana, back to the same geologic formation where he had found 104-A, back to a bone-hunting partner named Bob Makela whom he had known since the time in Missoula, and together these two aging hippies made a world-class paleontological discovery.

Throughout human history until the late eighteenth century, mankind had no inkling that any such beasts as the dinosaurs had ever existed. Only in 1841 did an Englishman coin the word *Dinosauria*, lumping certain strange new-found fossils into a category that translates as "terrible lizards." Actually they had been neither lizards (the lizards are a particular group of reptiles, distinct from both dinosaurs and crocodilians) nor, most of them, very terrible. Many were large herbivores, pacific creatures, making their livings in roughly the same way as a modern moose, or an elephant, or a giraffe. Even *Tyrannosaurus rex* may have been less the ferocious and implacable predator than commonly portrayed, more of a lazy and opportunistic omnivore, feeding on carrion or weakened animals or whatever was most convenient, as a grizzly bear does today. But for another 120 years, this remained the unshaken conventional view of dinosaurs, both in popular presentations (like Disney's *Fantasia*) and among the scientists: The meat-eating species were fierce predators that walked erect on hind legs, the vegetarians were huge gawkish vulnerable dolts, and all of them were simply magnified variations on the anatomy and physiology of a lizard. Cold-blooded. Mentally dim. Lacking any hint of advanced social behavior.

Finally a few scientists rebelled. That traditional view was not only unsupported by fossil evidence, they said; it was downright paradoxical.

In 1969 John Ostrom told a convention of paleontologists: "The evidence indicates that erect posture and locomotion probably are not possible without high metabolism and high uniform temperature." About the same time Armand de Ricqlès, a bone specialist in Paris, noticed that the internal structure of many dinosaur bones seemed to resemble mammal bones more closely than lizard bones. During the next several years Robert Bakker assembled a fuller framework of evidence that pointed the same way, and published a pair of revolutionary papers in the journal *Nature*. According to Bakker, the dinosaurs had been warm-blooded. Some of them, to help maintain their thermal stability, had developed an insulating layer of feathers. In their physiology, and most likely too in their behavior, they were advanced far beyond any lizard on Earth today. In fact, argued Bakker, they should not even be included among the reptiles. These animals had evolved into something distinct. Furthermore, said Bakker, "the dinosaurs never died out completely. One group still lives. We call them birds."

Following this line of thought, what Jack Horner and Bob Makela found on that Montana hillside was a great teeming dinosaur rookery. The first evidence of extended parental care, nesting in colonies, elaborate social interaction—those attributes linking dinosaurs with birds—that has ever been uncovered to human view.

Horner calls the site Egg Mountain, in a spirit of ironic but grateful homage. Actually it is only a gentle knoll, one among many out in this rolling terrain of sparse scrubby grass and hillocks and coulees cutting down through gray limestone. The real mountains loom up to the west, a towering wall of dark peaks and cliff faces not than a dozen miles off, snow-covered nine months of each year. That stretch of mountains, called the Sawtooth Range, is the easternmost front of the Rockies along this northern part of their length, the very juncture line where the great midland prairies come to a sudden halt, running smack up against the roof-beam of the continent. A few miles up the gravel road from Horner's Egg Mountain is another anomaly called Pine Butte Swamp, now protected by The Nature Conservancy because of its ecological uniqueness, a northern fenland of wolf willow and bog bean tucked flush against the base of the Rocky Mountain Front. The Pine Butte area is interesting to a biologist for forty reasons but noteworthy to any layman for one thing: It is the only place in the lower forty-eight states where *Ursus horribilis*, once the most formidable beast on the American prairies, still ventures out onto the plains.

"This is the last place in America," says Jack Horner, "that has the grizzly bear still in its original habitat. Out on the plains. Do you know what it's like to be on your knees, looking for dinosaur bones—and at the same time you have to look over your shoulder, watching for grizzly?" His face contorts into a lopsided smile. "It's exciting." Bears wander over occasionally from Pine Butte, to forage for roots or hunt rodents on the hillsides around Egg Mountain. "You come across a paw print, a *fresh* print, like eight inches long. And that land out there is just *open*. Nowhere to go. Not a tree to climb for miles." Another large grin.

Horner was on his knees like that, watching for small bones in the dirt and for big furry shapes over his shoulder, when he and Bob Makela made their historic discovery. On the side of Egg Mountain, in a bowl-shaped depression of brown mudstone, they found the skeletons of eleven baby dinosaurs of the Had-rosauridae family. The hadrosaurs were a group of semi-aquatic herbivores, also called "duck-billed" dinosaurs for the slightly comical shape of their plant-gathering jaws, and though adult hadrosaurs were well known from Montana and elsewhere, neither complete juvenile specimens nor eggs had ever been found. Close by the first eleven were another four skeletons of the same type and size.

The depression was unmistakably a nest. Patterns of deep wear on the teeth showed that these babies had been feeding, and for a longish period—yet here they were, in a crowded jumble, still clinging to the cradle. They seemed to have died from neglect; suddenly orphaned, perhaps, at an age when they weren't yet capable of going out to shift for themselves. In a paper published in *Nature*, Horner and Makela wrote: "The fact that 15 baby hadrosaurs had been feeding, and had stayed together for a period of time, indicates that some form of extended parental care was administered for, if the young were confined to the nest, food must have been brought to them." If so, those young hadrosaurs and their doting parents were unlike any reptiles known in the world today.

Horner and Makela described the new species and named it *Maiasaura peeblesorum*. The *peeblesorum* was in thanks to a family named Peebles, ranchers on whose land the find had been made. *Maiasaura*, according to Horner, means "good-mother reptile."

The excavations on Egg Mountain and in the surrounding area have continued for seven years, with no sign yet that this rich vein of fossils is even beginning to play out. More nests have turned up, more juveniles, and at least 300 whole or partial eggs. Adult *Maiasaura* have been found, as well as portions of adults from two other dinosaur species, one of which seems to have been a smaller carnivore, a swift creature that may have preyed upon young *Maiasaura*, snatching babies out of the nest when there was a lapse of parental protection. And along a certain ridge above Egg Mountain, stretching out for more than a mile, is what appears to be a continuous, staggeringly abundant deposit of hadrosaur bones. Three thousand pieces have already been taken from one little trench; by extrapolation, the entire ridge deposit might contain several million. That sheer volume of contemporaneous fossils suggests that a vast herd of hadrosaurs, hundreds of animals, once gathered here sociably in a huge clamorous breeding colony, a rookery, finding security in numbers for themselves and their nestlings. Much the way penguins do today.

In 1982 Horner left Princeton. He accepted a position as curator of paleontology at the Museum of the Rockies, a modest institution connected with Montana State University, in Bozeman. The salary is meager. The library resources at MSU are meager. The walls have no ivy. It's a museum where hadrosaur specimens share their end of the basement with a dry old Conestoga wagon. All of which is fine with Horner, who simply wanted to get back to Montana. The editors of *Nature*, in London, will not worry about his return address.

Through the winters—and out here they are long ones—he now studies his specimens, writes papers, teaches. Then in early June he moves north, with his tepee, to a camp site near Egg Mountain. In company with his old friend Makela (a science teacher at the only high school in Rudyard, Montana) and a few dozen assistants, volunteers, he digs and scratches at the ground. The camp's crucial field-season supplies include a rented jackhammer, short-handle picks, ice awls, delicate brushes, and 150 cases of beer. For three months, Horner is at large in the wild among *Maiasaura peeblesorum*, *Tyrannosaurus rex*, and the grizzly.

I asked Jack Horner if he could imagine any situation in life that he might prefer to the one he occupies now. He thought for a moment, carefully, and then said: "No."

# THE LIVES
# OF EUGÈNE MARAIS

## E Pluribus Unum
### in Termite and Man

A jellyfish is something much more than the sum of its parts, but that hasn't always been so. Early jellyfish ancestors followed simpler arithmetic. They were precisely, and only, the sum total of a grouping of similar cells; they were in fact colonies of individual unicellular animals, each individual not terribly different from an amoeba. In the primordial gumbo of Precambrian oceans, the consensus for togetherness came about first, then division of labor, finally morphological specializations that fitted certain of the individuals for certain tasks. Some members of the colony went into service as gut lining, some as tentacles. This pattern of evolution toward multicellular animals, which has not been uncommon in the origin of species—and which should be understood literally, not metaphorically—has been labeled *amalgamation*. The principle is as familiar as the print on a dime: *e pluribus unum*. Many simple lives fused synergistically into one complex life. Sponges evolved the same way. So did sea anemones and hydra and others of that gooey ilk. And so also, if we are to believe a wondrous lunatic named Eugène Marais, has an animal known as the *termitary*.

A termitary is a colony of termites.

In South Africa, where Eugène Marais spent most of his years, the predominant sort of termitary consists of sand particles heaped up like a giant pointy anthill, glandular secretions applied as mortar, fungus gardens in hidden damp compartments and the moving bodies of uncountable individual termites. There is also one termite queen, swollen grotesquely with ovulation, too fat to move, ensconced and well-tended within a royal chamber. Such a termitary might be forty feet high and hundreds of years old; it might include more than a million termites. A termitary found in the Limpopo valley, according to careful measurements and calculations made by an engineer friend of Marais, reportedly incorporated 11,750 tons of earth. Tons. And that mountainous pile of slobber-glued sand, with its intricate system of passages and rooms and ventilation ducts, with its hothouse mushroom patches, with its million living constituents clambering every-whichway, was in reality—so Marais argued—a single animal. He was quite serious.

Eugène Marais was born in 1872 near Pretoria, and within the space of sixty-four years he lived more lives than a Hindu cow. He was at various times a naturalist, a newspaper publisher, a lawyer, a journalist, a medical student, a smuggler of munitions, and one of the first important vernacular poets in the Afrikaans language. He was also a morphine addict and a suicide. Beyond this the details of his life—and not just the details, the basic facts, the dates—are little known, given vaguely and sometimes contra-dictorily in a very few sources. If Eugène Marais hadn't existed, it would have been Jorge Luis Borges who invented him. But he did exist; that little we know for sure. Because he left behind a matched pair of posthumous books just too bizarre to have been concocted.

These are *The Soul of the Ape* and *The Soul of the White Ant*. It is entirely typical of the warps and wobbles of factuality through-out the Eugène Marais story that the first of the two is not about apes, the second is not about ants, and that the pair were written in two different languages.

At the age of nineteen Marais was editor of a Pretoria newspaper called *Land en Volk*, and two years later he owned it. Pretoria in those days, just before the Boer War, was the capital of the Transvaal Republic and the site of its parliament, the Volksraad. Reporting and editorializing on parliamentary tomfoolery for his paper, Marais evidently was so scathing that, in the words of his son, "he had the distinction of being expressly excluded from the press gallery by a resolution of the Volksraad." Soon after that he stood up against Paul Kruger, the President and virtual dictator of the Transvaal, who was taking steps to repress public gatherings and the press, in mind of turning the Transvaal into an equatorial Prussia; suddenly, for his meddling, Eugène Marais was being tried on a charge of high treason. Before the Supreme Court at Pretoria, he beat that rap.

During this early period as a hard-charging political journalist, Marais was already showing the tendencies that would later make him a fanatically observant naturalist, of arcane sympathies. While running the newspaper, says his son, Marais had a great interest in animals and "was never without tame apes, snakes, scorpions, and the like." One of his favorite tame scorpions, according to Eugène Marais himself, was a badass female almost six inches long. This imposing creature once attacked and killed an adult chicken in ten minutes. But she would sit on Marais's hand, grip him kittenishly with her claws, hold back her sting. Marais wrote: "She liked being scratched gently."

In 1894 he married a young woman who bore the one son and then died. About this time, taking the loss very hard, Marais began his morphine habit. And the next year—in the first of his abrupt metamorphoses—he went to London, with the idea of studying law, or medicine, or both.

After four years of medical training—and this again is typical—he became a lawyer. Then the Boer War broke out; as an enemy national during wartime, he could only remain in London on parole status. He left. When the war ended in 1902, with the defeat of his people by the British, Marais was in Central Africa, preparing to smuggle a load of explosives and medical supplies

across the Limpopo River to embattled Boer forces. He had con-
tracted a bad case of malaria. The supplies were buried, and
Eugène Marais limped back to Pretoria. He settled in for a phase
of quietly practicing law.

Now the second metamorphosis: While living the life of a local
attorney, he emerged, almost magically, as one of South Africa's
most influential poets. The Afrikaans language was a slangy African hybrid of Dutch, still fresh and unrecognized in those days,
and Eugène Marais with his lyric poems seems to have done for
it something of what (in a much grander way) Dante did for
Italian. He showed that it was supple enough, beautiful enough,
to hold art. His poem "Winter Nag" has been called the heraldic
beginning of the new Afrikaans movement. Today in South Africa Marais is chiefly known, despite the two visionary books on
animal behavior, as an Afrikaans poet.

Naturally after a few years of lawyering he was bored and
disgusted. Metamorphosis number three: He retired to a remote
gorge in the mountainous Waterberg district, built himself a hut,
and lived there for three years in the company of a troop of 300
chacma baboons.

Long afterward he described that baboony time in a letter: "I
followed them on their daily excursions; slept among them; fed
them night and morning on mealies; learned to know each one
individually; taught them to trust and to love me—and also to
hate me so vehemently that my life was several times in danger.
So uncertain was their affection that I had always to go armed,
with a Mauser automatic under the left armpit like the American
gangster! But I learned the innermost secrets of their lives." Those
behavioral observations, and the innermost baboon secrets Marais
felt he had deduced, became the basis for a book which he hoped
would be his masterpiece, but which he never finished. A partial
manuscript was finally published, as *The Soul of the Ape*—though
not until 1969, and then only through the help of the late Robert
Ardrey, whose own *African Genesis* had been dedicated to Marais.

Marais harbored large ambitions about the ideas in that manuscript: "I have an entirely new explanation of the so-called sub-

conscious mind and the reason for its survival in man. I think that I can prove that Freud's entire conception is based on a fabric of fallacy." The kernel of his argument is that the human unconscious, as discovered and described by Freud, is nothing other than the older and more basic conscious mentality of the higher pre-human primates, which has been pushed into the psychological background, but not eliminated, by the newly evolved human consciousness. In other words, the human unconscious is identical with—in Marais's choice of phrase, and he meant it literally— the soul of the ape. A proposition which, to say the least, has few followers among modern psychologists.

The baboon-watch ended when Marais's recurrent malaria and his morphine habit (possibly also loneliness) drove him back again to Pretoria. But throughout his three years among the baboons and perhaps (the tapestry of known fact is at this point especially threadbare) for seven years thereafter, Marais was, in addition to all else, a passionate student of termites. He spent long hours watching them. He performed experiments. He traveled to inspect exceptional termitaries. He scratched his head and speculated. How, in a parched countryside, do a million termites satisfy their constant need for water? Why do they grow fungus and gather hay? How does the queen get from one royal chamber to another, when she is too large to fit through the door, incapable of moving herself, and too heavy to be lifted? For these mysteries and others Marais found solutions—some right, some wrong but plausible, some cockamamie. Above all he posed and answered the question: What manner of thing is a termitary? It is an organism, he said; a single living animal.

The life of a termitary begins with the nuptial flight, when a winged male and a winged female—each dispersed from an existing termitary—meet and mate and dig a small nest for their offspring. This founding pair, from which all the millions of other individuals will be directly descended, are in Marais's view the "generative organs" of the termitary. The king remains small while the queen grows hugely distended and is before long laying 50,000 eggs every twenty-four hours. The offspring are mainly wingless and asexual, divided into worker and soldier castes; these workers and soldiers, according to Marais, constitute respectively the red and white corpuscles of the bloodstream. Deep within the termitary, fungus gardens serve as stomach and liver, wherein vegetable food from outside is left to be decomposed by fungi before the termites themselves eat it. And meanwhile the queen in her hardened chamber, giving off a pheromonic essence that inspires purpose and cohesiveness among all the population, is (besides being the ovary—Marais's theory is not without some wooliness) the brain. Kill the queen and, true enough, the entire termitary dies.

Marais concluded: "The termitary is a separate composite animal at a certain stage of development, and lack of automobility alone differentiates it from other such animals. . . . [It is] an example of the method in which composite and highly developed animals like the mammals came into being." That is, by amalgamation.

Marais's termite studies appeared, beginning in 1923, as a series of short articles in various Afrikaans newspapers and in a magazine called *Die Huisgenoot*. A final and definitive article was published by *Die Huisgenoot* in 1925. Written in Afrikaans, it would have been intelligible only to Boers, Dutchmen and Flemings.

Maurice Maeterlinck, an eminent European playwright and Nobel Prize winner, happened to be a Fleming. Evidently he saw the 1925 article. The following year Maeterlinck published a book titled *The Life of the White Ant*, expropriating the detailed observations of Eugène Marais, and his terminology, and his theory about termite amalgamation. The book was a success throughout Europe. Marais got no acknowledgement. He hadn't the money to press a lawsuit.

Downward spirals of morphine and depression. Nine years passed, and then a woman in London, Winifred de Kok, began translating Marais's own termite pieces into English for their eventual publication as *The Soul of the White Ant*. She corresponded with Marais, and the letters to her tell us most of what we know about his inner life. He seemed to take new hope. He seemed buoyant for the first time in years. He was finally to have vindication. In one letter he wrote: "You see that your kindly enthusiasm has infected me! . . . The thought of reaching a bigger public intrigues me."

Five months later he put a shotgun to his head and fired.

What are we to make of such a man, such a life, such a set of lives? Robert Ardrey has called him "the purest genius that the natural sciences have seen in this century." Well, no. The works left behind by Eugène Marais simply cannot support that weight. What then?

As a newspaperman he ruined himself by excessive candor.

As a lawyer he was a good poet; or at least, an influential one. As a poet he was not Wallace Stevens. I once spent a day in the New York Public Library with several scarce old collections of his poems and (though I am admittedly a feeble judge of poetry) the stuff seemed to me pretty terrible. Maybe it was the translation. Maybe not.

As a descriptive naturalist he was, at his best, wonderful. But even *The Soul of the White Ant* is a badly flawed book, with patches of metaphysical gobbledygook and lame guesses and outright non sequiturs mixed in among the wonders.

As a theorist of insect and primate evolution he was, I think, more than a little brilliant and more than a little nuts.

What then is—as the reductionist jargon would have it—the bottom line? There is none. On Eugène Marais, no bottom line. He does not lend himself to categorization, easy dismissal, unreserved adulation, or summary assessment of any sort.

Except perhaps this. As an amalgamation of many individual lives, that polymathic phenomenon to which adhered the name Eugène Marais was one exotically complex jellyfish. He was a man of parts. But he was something much more than the sum of them.

# THE MAN
# WITH THE METAL
# NOSE

---

*Tycho Brahe and the
Olfactory Dimension of Science*

A very good friend of mine claims—among other matters of wild
and ornery personal ethic—that he chooses his friends strictly by
smell. I'm not sure what this says about my own aromaticity but
I do understand, and endorse, his point in principle: The ineffable
qualities are the ones that count, not the objective characteristics
that can be capsulized in an introduction or on a resumé or during
two hours' conversation over cocktails. Those ineffable qualities
will answer the more crucial questions upon which a friendship
is based, like *Would this person instinctively step between me and a
charging wart hog?* or *Could I trust him to borrow a book without
turning the page corners down?* And the nose, being humanity's most
underdeveloped sensory organ, is perhaps the only apt emblem
for our groping and sniffling efforts to register the ineffable. Which
is why I can't stop wondering about one particular nose that
occupies an intriguingly prominent place in the history of scientific inquiry.

It was an artificial one, this nose, a prosthesis made of gold
and silver alloy. It was worn by an aristocratic Danish astronomer

of the sixteenth century, a portly and sybaritic man named Tycho Brahe, who had lost his own God-given schnozzle in a duel. History does not record whether the replacement was held in position by a leather thong (as was that suspiciously similar one worn by Lee Marvin in *Cat Ballou*), or if not then how; but we do know that all his life Tycho carried a small snuffbox full of ointment, with which he constantly kept his metal nose lubricated, like one of those people compulsive about Chapstick. There is likewise no evidence as to what purpose this cold piece of technology might have served. The surviving portraits of Tycho suggest that its role was not to support eyeglasses. Did it smell? Did it run? Could it be turned up? Was it often out of joint? We'll never know. In fact Tycho himself, with or without his peculiar nose, would probably be forgotten completely by history if it weren't for two important considerations. The second of these was a set of notebooks full of numbers, and we'll come to that in a moment. The first was a galactic event of literally the greatest magnitude.

Step outside on a summer night and look off toward the northeastern part of the sky. Not far below the Little Dipper you'll see the constellation Cassiopeia, easily recognizable in the shape of a W. Back in early November of 1572, when Tycho Brahe was still a young amateur stargazer of twenty-five, a bright new star appeared suddenly in that constellation. It was shining with more brilliance than any other star, more brilliance than the planet Venus, so bright that it could be seen faintly even during daylight. Furthermore it was gleaming out from a spot where, just a week earlier and throughout the centuries before, no star at all had ever been visible. This phenomenon posed a serious philosophic problem in the late sixteenth century, when Aristotelean cosmology as sanctioned by the Catholic Church decreed that the upper celestial spheres—everything out there beyond the moon—were absolutely immutable. On the fourth day of Genesis, God had created the lights in the firmament, then left them alone, and that was that. Now suddenly here was a big brand new dot of fire flaunting its power in Cassiopeia. The star attracted attention, and concern, not just among theologians and astronomers. It was a popular event of mythic resonance. And it made the reputation of Tycho Brahe.

Tycho wasn't the first knowledgeable watcher to spot the new star, but he noticed it for himself one night before the news had gone public, and it left him agape. Over the next sixteen months, while the star changed color and rapidly dimmed, he performed a continuous sequence of very precise measurements, using a fine sextant he had crafted from walnut wood (the best available technology, the telescope not yet having been invented), that allowed him to speak about this new star more authoritatively than anyone else in Europe. It was immobile relative to Cassiopeia, Tycho stated; it was not in the sublunary atmosphere, but far beyond amid the other stars; it was not a comet without a tail, as some thought, but a true star. Tycho's book, *De Nova Stella*, made him internationally famous. He had charted all apparent aspects of the star with surpassing accuracy; but he had no idea what in the devil it was.

Today we know: It was a supernova explosion. Only five such cataclysmic events have been visible from Earth during the past thousand years, and of those, Tycho's in 1572 was the fourth. Some thousands of years earlier a gigantic star (much larger than our sun) had come to the end of its lifespan—the hydrogen nuclei at its core all "burnt" by fusion to form helium nuclei, and the helium further fused into still bigger nuclei. The star had then fallen into a terminal sequence of convulsions, alternately expanding and contracting, gravitational compaction seething down against rising internal pressures, which led to a final incredible thermonuclear explosion. That explosion flared out perhaps one billion times brighter than the star itself had shone. And the flash, having traveled across all those light-years between, eventually showed itself to Earth from the direction of Cassiopeia. Then by 1574 it was gone. No one knew why. Not even Tycho Brahe.

But Tycho, who had so faithfully measured and plotted the thing, was now a national hero in Denmark. The king gave him his own island, as well as lavish financial support with which to construct there a great astronomical observatory that would be Tycho's private scientific demesne. Tycho built a castle in Gothic Renaissance style, with spires and gables and cornices, and at the apex an onion dome topped by a gilt vane in the shape of Pegasus. There were guest rooms and aviaries and fountains, formal gardens and neat orchards laid out within a great perimeter wall, fish ponds, English mastiffs to stand guard, a paper mill, a print shop for his publications, and from ceiling to floor in the main workroom an oversized mural of Tycho himself. He called the place Uraniborg. The various chambers and towers he furnished with all the best astronomical instruments a king's money could order up: sextants of walnut, quadrants of brass and steel, armillary spheres ornamented with his own portrait, triquetrums and azimuth circles and astrolabiums—who knows what they all did. In this setting Tycho commanded his many assistants, threw grand parties for visiting nobility, rubbed ointment on his metal nose, and tossed scraps of meat to his attending dwarf Jeppe, who served as official court fool. Tycho, in other words, was not a scientist in the ascetic vein.

But during the next twenty years at Uraniborg he also performed the most precise and potentially useful collection of continuous astronomical observations that mankind up to that date had achieved. Where other astronomers (including most recently Copernicus) had been casual and sporadic about their own observations, Tycho was thorough, punctilious, indefatigable. Where others had tracked the planets with only their unaided eyes, occasionally a primitive sextant, Tycho devised his ingenious new instruments. Where others watched for a few nights or a few months, then went inside to dream up more or less misguided theories, Tycho kept watch relentlessly for over two decades, all the while recording his careful notes. The large quarto volumes containing those notes were his treasure. His contribution to science lay in recognizing that serious astronomy *required* data-gathering of such precision and continuity, and in marshalling the financial resources, the elaborate equipment, the patience, to make it possible. But again, as with the star of 1572, Tycho never knew what he had.

He was unpersuaded by the Copernican theory of celestial organization (which had been published quietly about fifty years earlier) and dissatisfied with the old Ptolemaic view. So in 1588 Tycho announced his own version. Earth, according to Tycho, was stationary in space, as Ptolemy had thought. The other planets, he said, moved in uniform circular motion around the sun. And the sun in turn orbited, pulling its satellites along, in a great graceful circle around Earth. This Tychonic system supposedly explained all the complex planetary motions that Tycho's sky-watching, over the years, had so accurately mapped. It was mathematically sweet and theologically acceptable. Its only drawback was that it was wrong.

After two decades at Uraniborg, where Tycho was a greedy and irresponsible landlord to the island's peasants, putting himself gradually into disfavor with the new Danish king, those munificent cash subsidies ended. So Tycho felt obliged to pack up his gear, his entourage, and leave. He went shopping across Europe for another royal patron willing to support him in similar high style, and two years later he found one: Tycho settled into a new castle just outside Prague, on the River Iser, under sponsorship of the Emperor Rudolph. Again there was money enough to pay for lordly living and a staff of assistants, among whom now was a twenty-nine-year-old German who had already earned modest recognition as an astronomer in his own right. This man's name was Johannes Kepler, and he had some ideas about celestial organization himself.

Kepler had abandoned everything to join Tycho in Prague, for a single ulterior reason: He hungered to see the data in those precious notebooks. But Tycho let him go hungry, assigning Kepler to some demeaning lesser chores, while refusing to share information with him as a colleague. Then in October of 1601, Tycho Brahe quite suddenly died. And Kepler got hold of the notebooks.

Within eight years, using Tycho's data, Johannes Kepler had formulated and published two laws that for the first time accurately explained the dynamics of our solar system, and thereby began the modern age in astronomy. The laws were as simple, once recognized, as they had been ineffable before. First, said Kepler, the planets (including Earth) travel around the sun not in circles but in ellipses, great oval orbits with the sun nearer one end. Second, each planet moves not at uniform speed but at a velocity that changes according to its distance from the sun. Today those statements might seem unexceptional. But in 1609, how many minds could have guessed that God would design a universe using *ovals* and *irregular motion?*

Something more was at work here than just astronomical training, hard thinking, and Tycho Brahe's data. What else? In many of the great scientific discoveries there seems to have been an additional mode of percipience that took up in the shadowy zone where pure rationality ended, a further faculty that helped point the way to the particular revolutionary idea. The word *intuition* is sometimes applied but, like a paper label on a bottle, only obscures what's inside. Arthur Koestler in his intriguing book on the early astronomers calls it "sleep-walking." Einstein spoke in his own case of "the gift of fantasy." As a young man of twenty-three, Isaac Newton suddenly glimpsed his law of gravity in little more time than an apple would take to fall from a tree. Alfred Russel Wallace came across natural selection with the same suddenness, supposedly during an attack of fever, after Darwin had labored over the problem for half a lifetime. Watson and Crick found the structure of DNA using tinkertoys, youthful cockiness, and another lab's x-ray photographs—photographs which until then had failed to be correctly interpreted.

In each of these entries upon the ineffable, something more was at work than mere cerebration.

And Kepler shaped his inherited Tychonic data into a vision of cosmological order that was ingeniously simple, drastically unorthodox, and true. But Tycho himself, evidently, just did not have the nose for it.

# VOICES
# IN THE WILDERNESS

*Francis Crick and Others*
*on the Germs from Outer Space*

Francis Crick is no common crank. Arguably he is no crank at all: This is the man who (with James Watson) discovered the structure of DNA and thereby won a share of the 1962 Nobel Prize, whose body of work over the past three decades places him high among the world's preeminent molecular biologists, who has even been acclaimed as second only to Charles Darwin in the history of British biology. He has always carried the reputation of a bold thinker, a careful scientist, a hard-headed skeptic. Lately, though, there is some cause for wondering whether Francis Crick might have stepped off the curb. It stems from a theory he calls Directed Panspermia.

Dr. Crick asks us to look favorably upon the suggestion that life on Earth originally arose, not from a touch by the out-stretched finger of God, not from a bolt of lightning stirring organic molecules in a primordial oceanic bouillon—but from single-celled organisms that arrived here, four billion years ago, in a spaceship.

No, they weren't doing the driving themselves. These first extraterrestrials envisioned by Crick would have been passive passengers, simple microbes on the order of bacteria or blue-green algae (both of which appear very early in the fossil record of earthly life, without any evident precursors) or perhaps even yeast. He posits that small samples of such creatures "traveled in the head of an unmanned spaceship sent to earth by a higher civilization which had developed elsewhere some billions of years ago. The spaceship was unmanned so that its range would be as great as possible. Life started here when these organisms were dropped into the primitive ocean and began to multiply." Mankind and the rest of our ecosystem have evolved, in short, from an episode of willful transgalactic infection.

Crick initially floated this notion of Directed Panspermia in 1973, in a brief scientific paper co-authored with the respected biochemist Leslie Orgel and published quietly in the journal *Icarus*. Then in 1981 Crick brought forth a book, titled *Life Itself*, in which he argued the case judiciously but determinedly, and at great length. It's safe to say that most of his biologist colleagues remained unpersuaded, and the more circumspect were no doubt embarrassed for him. Yet Crick did have some factual ammunition. He had arguments and probability calculations and syllogisms. Most important, he had a measure of crazy wild guts. Directed Panspermia may or may not be more probable, as a literal explanation for the origin of earthly life, than the first chapter of Genesis or the lightning-and-gumbo hypothesis. But the fact that it can be taken so seriously, by a world-renowned biologist, is itself an intriguing matter.

This theory is quite different from the Erich von Däniken stuff about ancient astronauts; it is more scientific, and it is older. The term *panspermia* (translatable to "seeds everywhere") evidently goes back to the pre-Socratic Greek philosopher Anaxagoras, who proclaimed that life had spread outward through the cosmos by way of seed-bearing matter spun off from a great central vortex. Espousing such ungodly ideas, Anaxagoras was put on trial for impiety by the wise civic elders of Athens, and spun off into exile himself. Thereafter his idea of panspermia was supplanted by a rival doctrine: spontaneous generation. This of course is the notion that living creatures can arise, fully formed and suddenly,

from inert matter. Aristotle among others was very big on spontaneous generation, claiming with his usual dogmatic certitude
that various insects erupt magically from vinegar and dung and
old wool, that frogs originate from the curdling of slime, that
mice could be born simply from damp earth. But if such little
miracles of genesis were possible in every dungpile and stagnant
well, there remained no logical need for those seeds blowing in
from across the cosmos. So for much of the next twenty centuries,
under Aristotle's far-reaching influence and the nearly universal
belief in spontaneous generation, panspermia was an idea in eclipse.

It did not begin gathering adherents again until a few sensible
eighteenth- and nineteenth-century scientists, most notably Louis
Pasteur, showed by persuasive experiments that spontaneous generation was a delusion. Maggots would *not* grow in old meat if it
was shielded under a bell jar. Germ cultures would *not* arise in
a sealed vial of milk after it had been boiled. Well then, if life
couldn't bring itself spontaneously into existence after all, where
had it originally come from? The answer was on file back in
Anaxagoras: Perhaps it came from Out There.

In 1821 a Frenchman named Sales-Guyon de Montlivault con-
jectured that seeds from the moon had accounted for the earliest
sprouting of earthly life. (Montlivault was writing thirty years
before Darwin but the idea that all life had evolved from a single
primitive beginning was, like those hypothetical seeds, in the air.)
During the 1860s H. E. Richter, a German, proposed that germs
were carried from one part of the universe to another aboard
meteorites; as evidence he cited the fact that some fallen meteorites
contained carbon, an element essential to all organic molecules.
Richter's meteorite hypothesis received a weighty endorsement
soon afterward, when Sir William Thomson (later and better
known as Lord Kelvin) declared in a formal address that "we must
regard it as probable in the highest degree that there are countless
seed-bearing meteoric stones moving about through space. If at
the present instant no life existed upon this earth, one such stone
falling upon it might, by what we blindly call *natural* causes, lead
to its becoming covered with vegetation." Notwithstanding Kel-

vin, certain important factors stood negative against the possibility of meteoric transport. These included the extreme cold of interstellar space (roughly −220°C.), which would kill most forms of microbial life known on Earth, and the matter of re-entry to a planetary atmosphere, during which even the hardiest little creature on the surface of a meteorite would be burned away like the heat-shield of an Apollo capsule.

Then in 1905 those difficulties were swept aside by a Swedish chemist named Svante Arrhenius, who proposed the first comprehensive and (for his time) scientifically sophisticated version of the theory of panspermia. Like Francis Crick, Svante Arrhenius was no common crank, having won *his* Nobel in 1903. Arrhenius ingeniously solved both problems—too much interstellar cold and too much re-entry heat—by suggesting a pair of new ideas: (1) that the actual space travelers had been, not bacteria in their more vulnerable adult form, but the *spores* of bacteria; and (2) that these very tiny spores moved across the galaxy, not as hitchhikers aboard meteorites, but solitarily, impelled by the physical pressure of starlight.

Spores are simply the durable reproductive bodies put out by creatures like bacteria and fungi. They are made for travel, resistant to cold and desiccation and even time, evolutionarily designed to bridge long periods of unfavorable environmental conditions. By the time Arrhenius wrote, bacterial spores had been shown to survive temperatures down to −252°C. Furthermore, they were minute enough to be lifted into an upper atmosphere on thermal currents, nudged off into empty space by electrical repulsion, and then pushed along through vast interstellar voids by radiation pressure, the sheer force of starlight acting like wind on a sailboat. Traveling in such a way from a distant star, bacterial spores could have reached Earth in a few hundred thousand years and then entered down through the atmosphere slowly, in gentle steps, without being incinerated. Falling on a hospitable medium, they could have burgeoned, multiplied, begun to evolve. That's the logic of panspermia as Arrhenius left it. And that's basically how it turned up again, seventy years later, in the opening three minutes of Phil Kaufman's 1978 film remake (not the 1956 classic) of *Invasion of the Body Snatchers*.

Unfortunately, neither Arrhenius nor Phil Kaufman allowed for ultraviolet radiation, which scientists now figure would kill any unshielded space travelers—even bacterial spores—long before they could cross the distance between stars.

That's why they must have been packaged, says Francis Crick, in a spaceship.

Crick offers design and performance specifications for that long-gone spaceship. He gives us informed speculations about the chemical and physical conditions on the planet from which it came. He calculates how many light-years of distance it might have had to cross, how long that might have taken, how the senders might have ensured survival of their microbial cultures for that length of eons. He shows with a flurry of large numbers that the universe is old enough for all this. He argues that microorganisms were a more probable choice as transgalactic missionaries of life—in preference to the advanced beings simply heading out themselves—because microorganisms with their compact durability could go much farther. In fact Crick supplies everything but a convincing motive. Why should those beings have *cared* how far they could spread their bacteria? Why should they *want* to mount such an elaborate effort toward infecting a distant planet with their germs, germs that might or might not evolve into forms as bizarre and monstrous as the giant squid, the bark scorpion, the Wall Street lawyer?

To that crucial point, Francis Crick offers no compelling answer.

But the editors of the journal *Icarus* evidently recognized the omission, and they came to our rescue. Immediately following "Directed Panspermia" by Crick and Orgel, in the same July 1973 issue, was a short paper by someone named John A. Ball, entitled "The Zoo Hypothesis." If John A. Ball had any scientific credentials whatsoever, he did not append them to his byline. It may have been his first scientific publication; it may have been his last. The abstract preceding his monograph reads:

> *Extraterrestrial intelligent life may be almost ubiquitous. The apparent failure of such life to interact with us may be understood in terms of the hypothesis that they have set us aside as part of a wilderness area or zoo.*

Of course—a wilderness area for the Betelgeusians, or the Tralfamadorians, or whomever. With all interstellar RVs prohibited, permits required, pack-out what you pack-in, and somewhere far off a transgalactic James Watt lurking malevolently behind a Darth Vader face. Will he or won't he grant leases to plunder and pillage? It all seems so perfectly and poetically logical, I'm surprised no one alerted us centuries sooner.

Let's hear it for John A. Ball. Let's hear it for the uncommon crank.

# ALIAS BENOWITZ SHOE REPAIR

*Heavy Metal, High Water,
and a Man in a Neoprene Suit*

I first heard about George Ochenski from a friend of mine who happens to be president of the Montana River-Snorkelers Association. We were in a fancy restaurant, as I recall, and there was wine involved. Ochenski had come to my friend's attention in the course of his (the friend's) presidential duties, which are in strict point of fact nonexistent. I should explain that the MRSA presidency is a purely honorary title, self-bestowed actually, because the MRSA is a mythical organization. This is all quite different, please note, from labeling the organization itself nonexistent. Certainly the Montana River-Snorkelers Association does exist (mainly over wine and beer at various bars and restaurants, occasionally also around a campfire); it just isn't *real*. An actual mythical entity, then, the MRSA, of roughly the same ontological status as the NCAA national championship in football, or the domino theory of international relations. You should look into this fellow Ochenski, my friend told me. He can be reached care of Benowitz Shoe Repair, in a tiny town called Southern Cross, up in the Flint Mountains above Anaconda. Have some more cabernet, I said. But sure enough it turned out to be true. Benowitz Shoe Repair is another mythical entity, existent in its own way but not real. George Ochenski is both mythical and real. Are you with me so far?

George Ochenski must certainly be the preeminent river-snorkeler in the Rocky Mountains. He has talent, commitment, infectious enthusiasm, broad experience, state-of-the-art equipment, and a measure of lunatic daring. He has precious little competition. Most important, he has self-abnegating dedication to a larger purpose.

Sometimes you have to snorkel a river, Ochenski believes, in order to save it.

So dedicated is George Ochenski, and so scornful of risk, that—if necessary to make a point—he is willing even to snorkel the Clark Fork River downstream from the Anaconda smelter.

Now a river-snorkeler (in case this isn't self-evident) is someone who swims downstream in a river with his face under water, enjoying the ride, watching the scenery, breathing through his little tube. A lazy, hypnotic pastime best practiced on pellucid trout streams in midsummer. A few of us have been toying at it for years.

But George Ochenski does not toy. He jimmies himself into a full wet suit, adds fins and a hood and neoprene gloves and a fanny pack holding three cans of beer, pulls a pair of skateboarding knee pads into place, defogs his mask, and jumps into rivers. Gentle rivers and raging whitewater monsters. Last year, for instance, he did thirty-eight miles of the Salmon in Idaho without benefit of a boat. Also last year, he leapt into the Quake Lake trench—an earthquake-contorted stretch of the Madison River famous for biting kayaks in half—and nearly died. On that run his mask was ripped off six times while he tumbled head over teakettle through a garden of sharp boulders; the trench, George admits today, was a miscalculation. In Montana this kind of behavior does not pass unnoticed. By word, and more discreetly by the looks on their faces, people frequently tell him: *Son, you must be out of your everlovin' skull.* But they said that to Orville Wright, and they were wrong. Then again, they said it to Evel Knievel, and they were right. George Ochenski figures somewhere in between.

He has an enduring though ambivalent attraction to what he himself classifies "death sports." Huge squinting grin from George as he acknowledges this ambivalence. Mountaineering. Ice-climbing. Scuba. Never a major injury, never a bad accident— unless you count the time he fell 600 feet down a steep rock slope in the Alaska Range and did a self-arrest on his nose. Back in those years he traveled exotically for serious climbing, with generous sponsorship from the equipment people, and took part in the first successful ascent of the west face of Alaska's Mt. Hayes. Scaled some breathtaking frozen waterfalls. Around the same time, a consummate autodidact, he turned himself into an expert cobbler, because he wasn't satisfied with the professional repair work on his climbing boots; before long he was doing work for his friends too, and they had rechristened him, whimsically and metonymically, "Benowitz Shoe Repair." Today he mostly stays close to the little wood-heated cabin at Southern Cross, in the front room of which stands a bass fiddle. The fiddle is a logical switch from tuba, which he played for thirteen years. Benowitz is a man of many skills.

Several years ago, in response to pressure both internal and external, he gave up the glorious climbing, thanked the sponsors, and settled down to being useful politically. He had come to feel that he owed something back to the mountains and rivers; meanwhile there happened to be a certain crisis brewing near home. He now makes his living as an editorial assistant to an author of textbooks on environmental science. The cabin is filled ceiling-high with an eclectic library. On one wall is a quote from Congressman Ron Dellums: "Democracy is not about being a damn spectator against the backdrop of tap-dancing politicians swinging in the winds of expediency." By disposition, George is certainly no spectator. Some people, particularly of the opposition, might still take him on first impression for a wild-haired, good-timing, reckless flake. They would be grievously mistaken. George Ochenski has an excellent brain, he has chutzpah, he has focus.

And in a small trailer up the hill behind his own cabin, where the ash from his cook stove can't fuddle its circuits, he has an Apple II computer, its floppy discs full of damning information concerning the Anaconda Minerals Company.

On September 29, 1980, the Anaconda Company announced that it was closing its copper-smelting operations at the town of Anaconda. This came as a severe shock to the 1,000 smelter workers suddenly unemployed, and marked the end of a century of awesome environmental pillage. For one hundred years the Company had cut down forests, poisoned streams, smelted copper, piled up vast mounds of slag, and filled the air of the county with a sulfurous smog, in exchange for the regular paychecks dispensed. Now the economics of copper had shifted. Goodbye, thanks for everything. "The Company thought they could just lock the doors and walk away," says George Ochenski.

He and a few other Anaconda folk, some of them former smelter workers, think otherwise. They are after the Company like a fice dog after a bear. They have formed an enraged-citizens' organization, pressured the governor, pressured the senators, pressured the EPA. They want more than goodbyes. They want reclamation. They want accountability. At very least they want precise information about the nature and magnitude of the poisonous mess left behind.

With sulfur dioxide no longer pouring from the smelter stack, the chief concern now is over toxic metals: lead, cadmium, mercury, zinc, copper itself, and especially arsenic. One hundred years of copper-smelting have left various concentrations of some or all of these in the waters, in the plants, in the soil, in the animals of the county. George Ochenski and his compatriots want to know: *How much?* How much was dumped in the ponds, how much was buried, how much is still blowing free off the smelter site? How much is already in our lungs and our bones? How much is ingested with each brown trout from the Clark Fork

River, if a person should be so lucky as to catch one of the surviving fish, and so foolhardy as to eat it?

How much lead? How much cadmium? How much arsenic? The Anaconda Company, no doubt, devoutly wishes that these questions would go away.

Sometimes you have to snorkel a river in order to save it. Guided by this dictum, George Ochenski loaded his gear into the back of my car. It was late in the season, Labor Day weekend, with the air already growing cool. We paused briefly, where the gravel lane down from Southern Cross joined the larger road, to check the Benowitz Shoe Repair mailbox. Then George led me off on a pair of brief but illuminating tours.

We went to the Big Hole River, across the Continental Divide from Anaconda and clear of the war zone over heavy metals. The Big Hole is still a pellucid trout stream. We jimmied ourselves into wet suits, added fins and hoods and neoprene gloves; I pulled George's one extra skate-boarding pad into position over my favorite knee. Masks were defogged, snorkels adjusted, and we jumped in.

The view was beautiful. Trout and whitefish looked me in the eye, aghast, and skittered away. Sculpins darted discreetly for cover. I observed the differences in underwater behavior among three different species of stonefly. I gazed at the funnel webs of *Arctopsyche* caddisfly larvae, down between rocks in the fast water, that I had read about often but never before seen. I found a mayfly nymph equipped with an elephantine pair of tusks. We passed through a few modest sets of rapids, where the current abruptly accelerated and the boulders came at me like blitzing linebackers who must be straight-armed away. After two hours of cruising we were nearly hypothermic, but the experience had been delightful.

Our second tour was to the Clark Fork River, downstream from the settling ponds into which the Anaconda Company has voided its years of industrial offal. "We're off to snorkel the Clark Fork," George told a friend as we pulled out of town. The friend looked puzzled. Huge squinting grin from George. "Then we'll come back and glow in the dark."

We snorkeled a long section of the Clark Fork. Here the water was turbid, visibility was poor. The rocks of the stream bed were largely cemented together with silt, leaving no habitat for stone-flies or *Arctopsyche*. I didn't see a single fish. I didn't see a single insect. Some people claim that the Clark Fork today is actually much improved over its sorry condition two decades ago, before the Company adopted certain technical measures to mitigate the toxicity of its releases. Maybe those people are right. But I remain skeptical. The river I was swimming through, with my eyes open and my nose very close to the bottom, was definitely no basis for passing out congratulations.

This dramatic lack of vitality proves nothing, of course, about what causal role the smelter wastes, and the erosion from denuded hillsides around Anaconda, may or may not still be playing. It simply correlates. Consider it, if you wish to, purest coincidence. It is not, however, mythical. It is real.

Later Benowitz and I were careful to shower ourselves down with clean water. "River-snorkeling" he told me, and he should know, "is not supposed to be a death sport."

# THE
# TREE PEOPLE

## A Man,
## A Storm, A Magnolia

Some humans have a special relationship with trees.

I'm thinking here not of the professional foresters, nor the academic dendrologists, certainly not the barrel-chested flannel-shirted fallers. No, it's gentler folk I have in mind. Persons neither scientific nor pragmatic, whose encounters with trees tend to be more intimate, more spontaneous, marked by an altogether different degree of sensitivity and—it might not be going too far to insert the word *mutual*—appreciation. People who can actually quiet themselves sufficiently to gaze at one individual tree and perceive there a real living creature conducting its own mortal business. This isn't so easy as it sounds. "The tree which moves some to tears of joy is in the eyes of others only a green thing which stands in the way," wrote William Blake. These genuine tree people are rare.

John Muir was one—read his account of riding out the thrills of a mountaintop storm while perched in the upper branches of a hundred-foot spruce. Another was that curious historical figure Jonathan Chapman, dead in 1845, later sentimentalized and Disneyfied under the name Johnny Appleseed. Still another is the British novelist John Fowles, who has written an interesting and little-known nonfiction book titled *The Tree*, in which he avows: "If I cherish trees beyond all personal (and perhaps rather peculiar) need and liking of them, it is because of this, their natural correspondence with the greener, more mysterious processes of mind—and because they also seem to me the best, most revealing messengers to us from all nature, the nearest its heart." Among this group of tree-loving people, too, is my own father.

Unlike Muir, he has never rambled solitarily among the California sequoias, my father. Unlike Fowles, he offers no elaborate philosophical theories for his special attachment. Like Chapman, he is simply a planter. Ever since I can remember—let's say at least thirty years—this man has been planting and tending and doting upon trees. He has never sold a board-foot of timber. He has never carried a bushel of fruit to a fair. He barely consents to own a saw. And behind him his life stretches out like a burgeoning, flourishing wood lot.

For a long time it made no particular sense to me.

At the beginning there was a half acre of nearly bare real estate, formerly farmland, at the suburban fringe of a city in the southern Midwest, not far from the Mason-Dixon line. Upon the half acre sat a new house and a stately old black walnut tree, not much else; beyond a fence marking the west edge was a wild stand of high grass and thistle, through which a foot trail led to unspoiled hardwood forest that went on for several miles. Across that fence was a miniature wilderness area for the delectation of young boys. Then, gradually, bits of the forest came to the half-acre lot.

A soft maple was planted up front near the road. A hard maple, just out the kitchen window. A sweet gum beside the driveway. A pin oak near the old well. An apple tree, off at the northwest corner, keeping company with the compost heap. A little dogwood. A Scotch pine, which often seemed to be struggling against heat prostration. More maples along that west fence. Eventually, getting fancy, a ginkgo. And a magnolia tree, a hapless and delicate magnolia, to the right of the front door. The earliest of these were poached as saplings from the adjacent woods, carefully trotted home on the foot path and replanted; later it became necessary to patronize a nursery. From my point of view (roughly waist-high then), the place had become a nursery itself. Finally the man in question bought me a rake, and I was not amused.

It had become necessary to patronize a nursery because, by the time I was old enough to operate that new rake, the wilderness area over the back fence had disappeared. Bulldozers had scraped it away. In its place there was now a tract suburb of medium-sized boxes. Paved driveways and sidewalks. Tulips. Myrtle in neat patches. Precious few trees.

I sat high in the crow's nest of that old black walnut tree, years passing, and watched this transformation. A valuable early lesson, with the resonance of a parable. Still grudging my time at the rake, I could see after a while that the man, the planter, my father, was not insane after all. Not even perverse. As the forest was massacred, as the neighborhood turned into concrete and crabgrass, on our small island there was a continual spreading of new branches. Local trees and exotics thrown together in strange juxtapositions, most of them thriving, fighting each other genially for sunlight and water and the attentions of the chief arborist. He was running his own gene bank. I would not want to be so high-flown as to call this half acre a *symphonic* orchestration of trees; but it represented, at least, a pretty good Dixieland band. The planter enjoyed his son's complete approbation for a couple more years, until an ice storm hit the magnolia, at which point it seemed that things were perhaps being taken too far.

This would have been about 1963. In that peculiar borderland climate, ice storms were a familiar enough (though not common) feature of what passed for winter. Cold sleet would begin falling at night, as the temperature dropped, and by morning the entire city would be glazed with a very beautiful and extremely treacherous eighth-of-an-inch layer of clear ice. Power lines down. Fender smashing against fender. Hips being fractured. The ice storm in 1963 was especially bad, and instead of an eighth-inch thickness there was a quarter-inch. Over any large surface area, like the crown of a tree, that amounted to considerable weight. The conifers, adapted to serious snows, could take it. The hardwoods were bare and streamlined. But the magnolia tree, a southern-bred creature, too trusting to drop its big trowel-shaped leaves from its brittle limbs after the autumn frosts, was caught in a wretched position.

Every leaf was lacquered thickly with ice, hundreds of pounds in all, the whole tree about to collapse like so much Steuben glass under a garbage compactor. And so, on that Saturday morning, there was the droll spectacle of a man and his fifteen-year-old son (the latter an unwilling draftee, with places to go and other enormously pressing things, now forgotten, to do) taking turns on a stepladder to break the ice—gingerly, one leaf at a time, with raw cold hands—off that desperate magnolia.

I remembered the magnolia's trauma, and its rescue, just yesterday while reading some scientific papers on an extraordinary species of tree called the bristlecone pine. The bristlecone was originally to be the main subject of this essay, since largely preempted, but never mind about that. The bristlecone will get its full starring role some other time. It can wait. It knows how. It is a tree accustomed to taking the long view. Unrecognized by biologists until about thirty years ago, bristlecone pines in the mountains of the American southwest are today known to be the oldest living creatures on Earth.

We're not talking about the age of a *species*, understand, but about venerable *individuals*. One noble specimen of bristlecone, found alive in 1964 on the shoulder of a high peak in eastern Nevada, was calculated to be 4,900 years old. That single tree sprouted from its seed and began putting out needles, in other words, around the same time the Egyptians established their first kingdom. No pyramids yet. The book of Genesis was still in galleys. This was a very old tree.

A number of curious facts emerged from those journal papers on the bristlecone—it is a species of paradoxical superlatives—most of which have no pertinence here, no bearing upon that image of the man, the ice, the stepladder, the magnolia. But two of them do.

First: Dendrologists have discovered that longevity, among individuals of this most long-lived of earthly creatures, is inversely related to the hospitableness of its living conditions. Age-wise, the tree thrives on adversity. The more harsh and ungiving its particular locale, the longer a bristlecone tends to live. At lower elevations within the mountains of its native range, places where soil is decent, wind and erosion are not extreme, water is available in good supply, the bristlecone grows large and robust—but does not seem to survive much beyond 1,500 years. (These ages can be gauged rather precisely, from a core sample or a full cross-section, by counting the annual rings laid down through the trunk.) At higher elevations, right up at the edge of the timberline, on steep south-facing slopes of stony soil that is poor in organic material and chemical nutrients, where little water is available, winds are relentless, growing conditions are generally lousy— here the bristlecone lives as a gnarled dwarf. But long. Often enough in this harshest environment a bristlecone survives its 4,000th birthday. Obvious moral: When the growing is tough, the tough keep growing.

The second odd fact is corollary to that one. Certain dendrologists who count bristlecone tree-rings have taken to writing of two separate characters of tree within the species. These are the "sensitive" bristlecones and the "complacent" bristlecones. The sensitive bristlecones are those that respond to climatic fluctuations, such as a year of exceptional drought, by laying down a drastically narrower growth ring that year, or possibly no ring at all. A complacent tree records no such response. Maybe it shouldn't be surprising that the complacent trees (as reported in *Science*, 1968) tend to be those that are younger and more comfortable. Meanwhile the ancient trees, struggling through 4,000 years of thirst and starvation, solitary on exposed ridges, grotesquely shaped, hunkered down, clinging to life with only one or two green branches—these are the ones that leave a record of sensitivity.

Now it seems to me that this discernment of "sensitivity" and "complacency" among various individual pine trees, silently living to millennial ages on their high mountain slopes, must constitute some sort of breakthrough percipience for the botanical sciences. And it compels me to wonder about that beleaguered magnolia on the half-acre wood lot at the edge of the Midwestern city.

Was the magnolia, by disposition, a complacent creature? Or was it, as I hope, sensitive? Did it *appreciate*, during the year of growth following that 1963 ice storm, what the man with the stepladder and the raw hands had done for it?

The house and the half acre have long since been sold to strangers. The magnolia, last time I drove past, looked neglected: a broken crown, whole branches on which the leaves were a sickly brown. Too bad. It may not survive another big ice storm. Certainly it will not live to see 4,000.

But does there perhaps remain, in its heartwood, some record—just a slight thickening to one annual ring, like a grateful sigh—of that mildly eccentric act of love? Can a magnolia remember a man?

# CAUSE
# FOR
# ALARUM

# JEREMY BENTHAM, THE *PIETÀ*, AND A PRECIOUS FEW GRAYLING

Rumor had it they were gone, or nearly gone, killed off in large numbers by dewatering and high temperatures during the bad drought of 1977. The last sizable population of *Thymallus arcti-cus*—Arctic grayling—indigenous to a river in the lower forty-eight states: *ppffft*. George Liknes, a graduate student in fisheries at Montana State University, was trying to do his master's degree on these besieged grayling of the upper Big Hole River in western Montana, and word passed that his collecting nets, in late summer of 1978, were coming up empty. The grayling were not where they had been, or if they were, Liknes for some reason wasn't finding them. None at all? "Well," said one worried state wildlife biologist, "precious few."

Grayling are not set up for solitude. Like the late lamented passenger pigeon, grayling are by nature and necessity gregarious, thriving best in rather crowded communities of their own kind. When the size of a population sinks below a certain unpredictable threshold, grayling are liable to disappear altogether, poof, evidently incapable of successful pairing and reproduction without the circumstantial advantage of teeming fellowship. This may have been what happened in Michigan. Native grayling were extinguished there, rather abruptly, during the 1930s.

The Michigan grayling and the Montana strain had been from time beyond memory the unique and isolated representatives of the species in temperate North America. They were glacial relicts, meaning they had gradually fled southward into open water during the last great freeze-up of the Pleistocene epoch; then, when the mile-thick flow of ice stopped just this side of the Canadian border and began melting back northward, they were left behind in Michigan and Montana as two separate pockets of grayling. These were trapped, as it turned out, cut off by hundreds of miles from what became the primary range of the species, across northern Canada and Alaska. They were stuck in warmish southern habitats overlapping the future range of dominance of *Homo sapiens;* their own future, consequently, insecure.

The Michigan grayling went first. They had been abundant in the upper part of Michigan's Lower Peninsula and in the Otter River of the Upper Peninsula. One report tells of four people catching 3,000 grayling in fourteen days from the Manistee River and hauling most of that catch off to Chicago. By 1935, not surprisingly, the Manistee was barren of grayling. Before long, so was the rest of the state. Sawlogs had been floated down rivers at spawning time, stream banks had been stripped of vegetation (causing water temperatures to rise), exotic competing fish had been introduced, and greedy pressure like that on the Manistee had continued. By 1940, the people of Michigan had just the grayling they were asking for: none.

In Montana, where things tend to happen more slowly, some remnant of the original grayling has endured—against similar adversities in less intense form—a bit longer. Even while disappearing during the past eighty years from parts of their Montana range, grayling have expanded into other new habitat. More accurately, they have been introduced to new habitat, in the zoological equivalent of forced school-busing: hatchery rearing and planting. As early as 1903, soon after the founding of the Fish Cultural Development Station in Bozeman, the state of Montana got into the business of manufacturing grayling; and for almost sixty years thereafter the planting of hatchery grayling was in great vogue.

The indigenous range of the Montana grayling was in the headwaters of the Missouri River above the Great Falls; they were well established in the Smith River, in the Sun River, and in the Madison, the Gallatin, and the Jefferson and their tributaries—notably, the Big Hole River. They had evolved as mainly a stream-dwelling species and existed in only a very few Montana lakes. However, they happened to be rather tolerant of low dissolved-oxygen levels, when those levels occurred in cold winter conditions (but not when the oxygen was driven out of solution by summer warming). This made them suitable for stocking in high lakes, where they could get through the winter on what minimal oxygen remained under the ice. In 1909, 50,000 grayling from the Bozeman hatchery were planted in Georgetown Lake. Just a dozen years later, 28 million grayling eggs were collected from Georgetown, to supply hatchery brood for planting elsewhere. And the planting continued: Ennis Lake, Rogers Lake, Mussig-brod Lake, Grebe Lake in Yellowstone National Park. Between 1928 and 1977, millions more grayling were dumped into George-town Lake.

Unfortunately, that wasn't all. Back in 1909, hatchery grayling were also planted in the Bitterroot and Flathead rivers, on the west side of the Continental Divide, in stream waters they had never colonized during their ancestral migration. An innocent experiment, and without large consequences, since the grayling introduced there evidently did not take hold. But then, in what may have seemed a logical extension of all this hatchery rearing and planting, the Big Hole River was planted with grayling. The Big Hole already had a healthy reproducing population of wild grayling, but that was not judged to be reason against adding more. From 1937 until 1962, according to the records of the Montana Department of Fish, Wildlife, and Parks (FWP), more than five million grayling from the Anaconda hatchery were poured into the Big Hole, from the town of Divide upstream to the headwaters: hothouse grayling raining down on wild grayling.

This was before FWP biologists had come upon the belated realization that massive planting of hatchery fish in a habitat where the same species exists as a reproducing population is the best of all ways to make life miserable for the wild fish. Things are done differently these days, but the mistake was irreversible. The ambitious sequence of plantings was very likely the most disastrous single thing that ever happened to the indigenous grayling of the Big Hole.

At best, each planting instantaneously created tenement conditions of habitat and famine conditions of food supply. In each place where the hatchery truck stopped, the river became a grayling ghetto. At worst, if any of the planted fish survived long enough to breed with each other and interbreed with the wild fish, the whole planting program served to degrade the gene pool of the Big Hole grayling, making them less capable of surviving the natural adversities—drought, flood, temperature fluctuation, predation—of their natural habitat.

But here's the good news: Very few of those planted grayling would have survived long enough to breed. The mortality rate on hatchery grayling planted in rivers is close to 100 percent during the first year, and most don't last even three months, whether or not they are caught by a fisherman. These planted grayling come, after all, from a small sample of lake-dwelling parents, with little genetic variety or inherited capacity for coping with moving water. Reared in the Orwellian circumstances of the hatchery, cooped in concrete troughs, without a beaver or a merganser to harry them, eating Purina trout chow from the hand of man, what chance have they finally in the most challenging of habitats, a mountain river? The term "fish planting" itself is a gross misnomer, when applied to dropping grayling or trout into rivers; there is no delusion, even among the hatchery people, that these plants will ever take root. More realistically, it's like providing an Easter egg hunt for tourists with fishing rods.

In 1962 the Big Hole planting ceased, and the remaining wild grayling, those that hadn't died during the famine and tenement periods, were left to get on as best they could. Then came the 1977 drought and, the following year, the George Liknes study. One of Liknes's study sections on the Big Hole was a two-mile stretch downstream from the town of Wisdom to just above the Squaw Creek bridge. On a certain remote part of the stretch a rancher had sunk a string of old car bodies to hold his hayfield in place. From that two-mile stretch, using electroshocking collection equipment that is generally reliable, Liknes did not take a single grayling. This came as worrisome news to me because, on a morning in late summer of 1975, standing waist deep within sight of the same string of car bodies and offering no great demonstration of angling skill, I had caught and released thirty-one grayling in four hours. Now they were either gone or in hiding.

Grayling belong to the salmonid family, as cousins of trout and whitefish. In many ways they seem a form intermediate between those two genera; in other ways, they depart uniquely from the salmonid pattern.

The first thing usually noted about them, their distinguishing character, is the large and beautiful dorsal fin. It sweeps backward twice the length of a trout's, fanning out finally into a trailing lobe, and it is, under certain specific conditions, the most exquisitely colorful bit of living matter to be found in the state of Montana; spackled with rows of bright turquoise spots that blend variously to aquamarine and reddish-orange toward the front of the fin, a deep hazy shading of iridescent mauve overall, and along the upper edge, in some individuals, a streak of shocking rose. That's in the wild, or even stuck on a hook several inches underwater. Lift the fish into air, and it all disappears. The bright spots and iridescence drain away instantaneously, the dorsal folds down to nothing, and you are holding a drab gun-metal creature that looks very much like a whitefish. The grayling magic vanishes, like a dreamed sibyl, when you pull it to you.

Except for this dorsal fin, the grayling does resemble that most maligned and misunderstood of Montana fish, its near relative, the mountain whitefish. Both are upholstered—unlike the trout— with large stiff scales, scales you wouldn't want to eat. Both have dull-colored bodies, grayish-silver in the grayling, brownish-silver in the whitefish—though the grayling is marked along its forward flank with another smattering of spots, these purplish-black, playing dimly off the themes in the dorsal. They are also distinguishable (from each other and from their common salmonid relatives) by the shape of the mouth. A trout has a wide, sweeping, toothy grin; a whitefish's mouth is narrow and toothless—worse, it is set in a snout that is pointed and cartilaginous, like a rat's, probably the main single cause of the whitefish's image problem. The grayling, as you can see if you look closely, has been burdened with a mouth that is an uneasy compromise between the two: The narrow mouth is set with numerous tiny teeth and fendered with large cartilaginous maxillaries, but its shortened nose couldn't fairly be called a snout. The point is this: The grayling is one of America's most beautiful fish, but only a few subtle anatomical strokes distinguish it from one of the most ugly. A lesson in hubris.

But a superfluous lesson, since the grayling by character is anything but overweening. It is dainty and fragile and relatively submissive. With tiny teeth and little moxie, it fails in all territorial competition against trout—and this is another reason for its decline in the Big Hole, where rainbow and brown and brook trout that have been moved into the neighborhood now bully it mercilessly. Like many beautiful creatures that have known fleeting success, it is dumb. It seeks security in gregariousness and these days is liable to find, instead, carnage. When insect food is on the water, and the fish are attuned to that fact, a fisherman can stand in one spot, literally without moving his feet, and catch a dozen grayling. Trout are not so foolish: Drag one from a hole and the word will be out to the others. The grayling cannot take such a hint. In the matter of food it is an unshakable optimist; the distinction between a mayfly on the water's surface and a hook decorated with feathers and floss is lost on it. But this rashness, in the Big Hole for example, might again be partly a consequence (as well as a cause) of its beleaguered circumstances. The exotic trouts, being dominant, seize the choice territorial positions of habitat, and the grayling, pushed off into marginal water where a fish can only with difficulty make a living, may be forced to feed much more recklessly than it otherwise would.

At certain moments the grayling seems even a bit stoic, as though it had seen its own future and made adjustments. This is noticeable from the point of view of the fisherman. A rainbow trout with a hook jerked snug in its mouth will leap as though it were angry, furious—leap maybe five or six times, thrashing the air convulsively each time. If large, it will run upstream, finally to go to the bottom and begin scrabbling its head in the rubble to scrape out the hook. A whitefish, unimaginative and implacable, will usually not jump, will never run, will stay near the bottom and resist with pure loutish muscle. A grayling will jump once if at all and remain limp in the air, leaping the way a Victorian matron would faint into someone's arms—with demure, trusting abdication. Then, possibly after a polite tussle, the grayling will let its head be pulled above the water's surface, turn passively onto its side, and allow itself to be hauled in. Once beaten, a rainbow can be coaxed with certain tricks of handling to give you three seconds of docility while you get the hook out to release it. A whitefish will struggle like a hysterical pig no matter what. A grayling will simply lie in your hand, pliant and fatalistic, beautiful, placing itself at your mercy.

So no one has much use for the grayling, not even fishermen. It grows slowly, never as large as a trout, and gives unsatisfactory battle. It is scaly, bony, and not especially good to eat. Montana's fish and game laws will allow you to kill five of them from the Big Hole River in a day,* and five more every day all summer— but what will you do with them? Last year a Butte man returned from a weekend on the river and offered a friend of mine ten

*At the time this essay appeared in *Audubon*; since then, the regulations have been changed in the grayling's favor.

grayling to feed his cat. The man had killed them because he caught them, very simple logic, but then realized he had no use for them. This year my friend's cat is dead, through no fault of the grayling, so even that constituency is gone. A grayling does not cook up well, it does not fight well. It happens to have an extravagant dorsal fin, but no one knows why. If you kill one to hang on your wall, its colors will wilt away heart-breakingly, and the taxidermist will hand you back a whitefish in rouge and eye shadow. The grayling, face it, is useless. Like the auk, like the zebra swallowtail, like Angkor Wat.

In June of 1978, the U.S. Supreme Court ruled that completion of the Tellico Dam on the Little Tennessee River was prohibited by law, namely the 1973 Endangered Species Act, because the dam would destroy the only known habitat of the snail darter, a small species of perch. One argument in support of this prohibition, perhaps the crucial argument, was that the snail darter's genes might at some time in the future prove useful—even invaluable—to the balance of life on Earth, possibly even directly to humanity. If the *Penicillium* fungus had gone extinct when the dodo bird did, according to this argument, many thousands of additional human beings by now would have died of diphtheria and pneumonia. You could never foresee what you might need, what might prove useful in the line of genetic options, so nothing at all should be squandered, nothing relinquished. Thus it was reasoned on behalf of snail darter preservation (and thus I have reasoned elsewhere myself). The logic is as solid as it is dangerous.

The whole argument by utility may be one of the most dangerous, even ominous, strategic errors that the environmental movement has made. The best reason for saving the snail darter was this: precisely because it is flat useless. That's what makes it special. It wasn't put there, in the Little Tennessee River; it has no ironclad reason for being there; it is simply there. A hydroelectric dam, which can be built in a mere ten years for a mere $119 million, will have utility on its side of the balance against snail darter genes, if not now then at some future time, when the cost of electricity has risen above the cost of recreating the snail darter through genetic engineering. A snail darter arrived at the hard way, the Darwinian way, across millions of years of randomness, reaching its culmination as a small ugly perch roughly resembling an undernourished tadpole, is something far more precious than a net asset in potential utility. What then, exactly? That isn't easy to say, without gibbering in transcendental tones. But something more than a floppy disc storing coded genetic lingo for a rainy day.

Another example: On a Sunday in May, 1972, an addled Hungarian named Laszlo Toth jumped a railing in St. Peter's Basilica and took a hammer to Michelangelo's *Pietà*, knocking the nose off the figure of Mary, and part of her lowered eyelid, and her right arm at the elbow. The world groaned. Italian officials charged Toth with crimes worth a maximum total of nine years' imprisonment. Some people, but no one of liberal disposition, said aloud at the time that capital punishment would be more appropriate. In fact, what probably should have been done was to let Italian police sergeants take Toth out into a Roman alley and smack his nose off, and part of his eyelid, and his arm at the elbow, with a hammer. The *Pietà* was at that time 473 years old, the only signed sculpture by the greatest sculptor in human history. I don't know whether Laszlo Toth served the full nine years, but very likely not. Deoclecio Redig de Campos, from the Vatican art-restoration laboratories, said at the time that restoring the sculpture, with glue and stucco and substitute bits of marble, would be "an awesome task that might take three years," but later he cheered up some and amended that to "a matter of months." You and I know better. The Michelangelo *Pietà* is gone. The Michelangelo/de Campos *Pietà* is the one now back on display. There is a large difference. What, exactly, is the difference? Again hard to say, but it has much to do with the snail darter.

Sage editorialists wrote that Toth's vandalism was viewed by some as an act of leftist political symbolism: "Esthetics must bow to social change, even if in the process the beautiful must be destroyed, as in Paris during *les èvènements*, when students scrawled across paintings 'No More Masterpieces.' So long as human beings do not eat, we must break up ecclesiastical plate and buy bread." The balance of utility had tipped. The only directly useful form of art, after all, is that which we call pornography.

Still another example: In May of 1945 the Target Committee of scientists and ordnance experts from the Manhattan Project met to hash out a list of the best potential Japanese targets for the American atomic bomb. At the top of the list they placed Kyoto, the ancient capital city of Japan, for eleven centuries the source of all that was beautiful in Japanese civilization, the site of many sacred and gorgeous Shinto shrines. When he saw this, Henry L. Stimson, a stubbornly humane old man who had served as Secretary of State under Herbert Hoover and was now Truman's inherited Secretary of War, got his back up: "This is one time I'm going to be the final deciding authority. Nobody's going to tell me what to do on this. On this matter I am the kingpin." And he struck the city of shrines off the list. Truman concurred. Think what you will about the subsequent bombing of Hiroshima—unspeakably barbarous act, most justifiable act in the given circumstances, possibly both—think what you will about that; still the sparing of Kyoto, acknowledged as a superior target in military terms, was very likely the most courageous and imaginative decision anyone ever talked Harry Truman into. In May of 1945, the shrines of Kyoto did not enjoy the balance of utility.

"By utility is meant that property in any object, whereby it tends to produce benefit, advantage, pleasure, good, or happiness (all this in the present case comes to the same thing), or (what comes again to the same thing) to prevent the happening of mischief, pain, evil, or unhappiness to the party whose interest is considered: if that party be the community in general, then the happiness of the community; if a particular individual, then the happiness of that individual." This was written by Jeremy Bentham, the English legal scholar of the eighteenth century who was a founder of that school of philosophy known as utilitarianism. He also wrote, in *Principles of Morals and Legislation*, that "an action then may be said to be conformable to the principle of utility . . . when the tendency it has to augment the happiness of the community is greater than any it has to diminish it." In more familiar words, moral tenets and legislation should always be such as to achieve the greatest good for the greatest number. And *the greatest number* has generally been taken to mean (though Bentham himself might not have agreed: see "Animal Rights and Beyond") the greatest number of *humans*.

This is a nefariously sensible philosophy. If it had been adhered to strictly throughout the world since Bentham enunciated it, there would now be no ecclesiastical plate or jeweled papal chalices, no Peacock Throne (vacated or otherwise) of Iran, no Apollo moon landings, no Kyoto. Had it been retroactive, there would be no Egyptian pyramids, no Taj Mahal, no texts of Plato; nor would there have been any amassing of wealth by Florentine oligarchs and hence no Italian Renaissance; finally, therefore, no *Pietà*, not even a mangled one. And if Bentham's principle of utility—in its economic formulation, or in thermodynamic terms, or even in biomedical ones—is applied today and tomorrow as the ultimate touchstone for matters of legislation, let alone morals, then there will eventually be, as soon as the balance tips, no snail darter and no. . . .

But we were talking about the Big Hole grayling. George Liknes was finding few, and none at all near the string of car bodies, and this worried me. I had some strong personal feelings toward the grayling of the Big Hole—proprietary is not the right word, too presumptuous; rather, feelings somewhere between cherishing and reliance. I had come to count on the fact, for cheer and solace in a very slight way, that they were there, that they existed—beautiful, dumb, and useless—in the upper reaches of that particular river. It happened because I had gone up there each year for a number of years—usually in late August, which is the start of autumn in the upper Big Hole Valley, or in early September—with two hulking Irishmen, brothers. Each year, stealing two days for this pilgrimage just as the first cottonwoods were taking on patches of yellow, we three visited the grayling.

At that time of year the Big Hole grayling are feeding, mainly in the mornings, on a plague of tiny dark mayflies known as *Tricorythodes* (or, for convenience, trikes). One of these creatures is roughly the size of a caraway seed, black-bodied with pale milky wings; but they appear on the water by the millions, and the grayling line up in certain areas to sip at them. The trike hatch happens every August and September, beginning each morning when the sun begins warming the water, continuing daily for more than a month, and it is one of the reasons thirty-one grayling can be caught in a few hours. The trike hatch was built into my understanding with the Irishmen, an integral part of the yearly ritual. Trike time, time to visit the Big Hole grayling.

Not stalk, not confront, certainly not kill and eat; visit. No great angling thrills attach to catching grayling. You don't fish at them for the satisfaction of fooling a crafty animal on its own terms, or fighting a wild little teakettle battle handicapped across a fine leader, as you do with trout. The whole context of expectations and rewards is different. You catch grayling to visit them: to hold one carefully in the water, hook freed, dorsal flaring, and gape at the colors, and then watch as it dashes away. This is good for a person, though it could never be the greatest good for the greatest number. I had visited them regularly at trike time with the two Irishmen, including the autumn of the younger brother's divorce, and during the days just before the birth of the older brother's first daughter, and through some weather of my own. So I did not want to hear about a Big Hole River that was empty of grayling.

A fair question to the Montana Department of Fish, Wildlife, and Parks is this: If these fish constitute a unique and historic population, a wonderful zoological rarity within the lower forty-eight states, why let a person kill five in a day for cat food? FWP biologists have offered three standard answers: (1) Until George Liknes finished his master's thesis, they possessed no reliable data on the Big Hole grayling, and they do not like to make changes in management procedures except on the basis of data; (2) grayling are very fecund—a female will sometimes lay more than 10,000 eggs—and so availability of habitat and infant mortality and competition with trout are the limiting factors, not fishing pressure; and (3) these grayling are glacial relicts, meaning they have been naturally doomed to elimination from this habitat, and mankind is only accelerating that inevitability.

Yet, (1) over a period of twenty-five years, evidently without the basic data that would have showed that it was all counter-productive, the department spent a large pile of money to burden the Big Hole grayling with five million hatchery outsiders; and (2) though fishermen are admittedly not the limiting factor on total number of grayling in the river, they can easily affect the number of large, successful, genetically gifted spawning stock in the population, since those are precisely the individual fish that fishermen, unlike high temperatures or low oxygen or competitive trout, kill in disproportionate number. There might be money for more vigorous pursuit of data, there might be support for protecting the grayling from cats, but the critical constituency involved here is fishermen, and the balance of utility is not on the side of the grayling; as for (3), not only are the rivers of Montana growing warmer with the end of the Ice Age, but the Earth generally is warming; it is in fact falling inexorably into the sun, and the sun itself is meanwhile dying. So all wildlife on the planet is doomed to eventual elimination, and mankind is only et cetera.

The year before last, the Irishmen and I missed our visit: The older brother had a second daughter coming, and the younger brother was in Germany, in the Army, soon to have a second wife. I could have gone alone but I didn't. So all I knew of *Thymallus arcticus* on the upper Big Hole was what I heard from George Liknes: not good. Through the winter I asked FWP biologists for news of the Big Hole grayling: not good.

Then one day in late August last year, I sneaked away and drove up the Big Hole toward the town of Wisdom, specifically for a visit. I stopped when I saw a promising arrangement of water, a spot I had never fished or even noticed before, though it wasn't too far from the string of car bodies. I didn't know what I would find, if anything. On the third cast I made contact with a twelve-inch grayling, largish for the Big Hole within my memory. Between sun-on-the-water and noon, using a small fly resembling a *Tricorythodes*, I caught and released as many grayling as ever. As many as I needed.

I could tell you where to look for them, I could suggest how you might fish for them, but that's not the point here. You can find them yourself if you need to. Likewise, it's tempting to suggest where you might send letters, whom you might hector, what pressures you might apply on behalf of these useless fish; also not exactly the point. I merely wanted to let you know: They are there.

Irishmen, the grayling are still there, yes. Please listen, the rest of you: They are there, the Big Hole grayling. At least for now.

# HOMAGE
# TO BANGI BHALE

*The Bengal Tiger
in the Mind's Eye*

You can see a tiger in a zoo. You can stroll up within a few yards
and gawk at the big cat while, on the far side of the bars, it lies
sunk in a dispirited afternoon torpor or paces restlessly back and
forth. You can admire the gorgeous fur and the musculature and
the huge soft paws at close range; tap on the Plexiglas for a
reaction; pose yourself in the foreground for a smug snapshot.
Zoo tigers are available nowadays—like croissant sandwiches and
electronic banking—in every major civilized city. So what's the
point of flying halfway around the world for a short glimpse of
one of these creatures, by moonlight, at eighty yards distance,
through a pair of binoculars?

That's what I asked myself as I trudged back down the path
after a three-minute audience. I was a guest on this journey, but
still. I had flown 10,000 miles out to Nepal, ridden another hundred
or so in Land Rovers and dugout canoes and on the back of an
elephant, then begun walking and finally—just the last 200 yards—
abandoned even my shoes, as instructed, to proceed quietly up
the path in stocking feet. All for a brief look at a wild tiger. A
large male with blood-darkened forelegs, as it turned out, who
at that particular moment of that particular evening happened to

be noshing the hind end off a young domestic buffalo, which he had killed with a neck bite not many minutes before. This tiger's name, according to the local Nepali trackers, was Bangi Bhale: male with a crooked paw. He was far away down the gully slope and I was peering out through a small window in an elaborate elephant-grass blind. Around me other tourists and pilgrims were doing likewise. The borrowed binoculars helped, but still my view of Bangi Bhale was tantalizingly distant and dim. A peep show. More vivid were the loud ringing sounds of cracking bone, as he ate his way forward along the buffalo's rib cage.

For detailed observations of tiger anatomy, I could have done better at a zoo. For behavioral insight, the books by George Schaller and Charles McDougal offer vastly more. For a beautiful tableau, a study in sheer color and grace, the best of the glossy photographs are superior. But never mind. My head was already full of information and theories and frozen images; the purpose here now was different. Why had I stumbled out—for just a shamefully brief, dilettantish visit—to this ravine of this forest in this little country? Not with the expectation of learning much. Nor even of seeing much. But it seemed worthwhile, if only once in a life, to be at least *in the presence* of a wild tiger. I had come to pay homage.

The tiger is, arguably, the most formidable land creature on Earth. Without question it is the largest of all the big cats, and the dominant predator of Asia. It preys on a wide variety of mammals, from porcupine and monkey up to the heftiest wild buffalo; keeps its distance prudently from elephant and adult rhino; and is seriously threatened by no other animal except the dreaded *Homo sapiens*. Man is its ultimate enemy precisely because the tiger, like man, is smart, cautious, and highly skilled at killing.

There is only one tiger species, *Panthera tigris*, and it is unique to Asia. Some paleontologists suspect that it evolved in the north part of the continent, spreading south during the most recent epoch into the Caspian Sea area, the Indian subcontinent, southern China, and Indochina. Others interpret the fossil evidence to suggest an opposite view: that the tiger might have appeared first in Southeast Asia, radiating north and west along a great fan. It can still be found in the far eastern USSR, where it tends to be larger and more pale than in southerly zones. This broad distribution reflects the tiger's flexibility in adjusting to a spectrum of habitats, from snowy forest with temperatures at minus 30, to steamy tropical jungle. Scientists have traditionally recognized eight geographical subspecies, of which four (the Caspian tiger, the Javan, the Chinese, the Balinese) seem to be already extinct. Only in Siberia, in Sumatra, in Indochina, and on the Indian subcontinent does the tiger seem to have some chance of surviving in sizable, wild populations.

But even in these places the species is drastically reduced. Fifty years ago, a plausible estimate put the total number for all eight subspecies at around 100,000. In 1940, another rough guess gave just the Bengal subspecies (that race which inhabits the Indian subcontinent) a population of 40,000. This didn't sound bad; probably even back then it was far too optimistic. Hunting pressure had been heavy for more than a century, and those great fashionable tiger hunts (staged by British adventurers, Indian maharajahs, Nepali princes) were often so remorselessly efficient, with circled elephants and human beaters driving the cats toward the guns, that they amounted to systematic slaughter. During one hunt in Nepal, just after World War I, 120 tigers were killed. An Indian maharajah wrote to George Schaller as late as 1965 that "My total bag of Tigers is 1150 (one thousand one hundred and fifty only)." But the tiger has a high reproductive potential, and even hunting so profligate as this was not the worst of its problems.

Then the world discovered DDT, and the tiger in India and Nepal began losing its habitat—habitat which until then had been guarded by malarial mosquitoes.

With the big mosquito-eradication campaigns of the 1950s, with improved public-health measures against malaria, came the first chance for impoverished and overcrowded humans to settle in those lowland jungles. Of course the humans brought with them axes, machetes, plows, cattle, goats—all very effective at turning tiger habitat into sparse, open farmland. By the end of the 1960s, a careful census throughout India counted just 1,800 tigers. And Nepal could claim only about 150.

At this point a handful of very senior conservationists—from the Bombay Natural History Society, the World Wildlife Fund, and the International Union for the Conservation of Nature and Natural Resources—raised the alarm, instigating a global effort to rescue the tiger from annihilation. Initially, that effort was focused on the Bengal subspecies. Money was raised, scientific assistance was offered, fur boycotts were organized, international trade covenants were ratified; tactful appeals were made to high public officials of India, Nepal, and Bangladesh. Each of those countries takes some pride in the tiger as a national symbol. For another incentive, there was the wild tiger's potential as an attraction for tourists, and thus as a significant source of foreign currency. To the credit of the people and the governments of all three countries, the responses were positive and quick.

Laws were passed that banned the hunting of tigers. Enforcement measures against poaching were established. Equally crucial—and politically more difficult in these crowded developing nations—parcels of prime habitat were set aside as wildlife reserves. Those areas don't guarantee that *Panthera tigris* will be able to survive in the wild, but they vastly improve the possibility. India now has a dozen well-managed reserves that protect the tiger and its whole constellation of prey. Bangladesh has three, in the Sunderbans region at the mouth of the Ganges. And Nepal has three, including a 360-square-mile tract of grassy bottomland and deciduous forest, just south of the Himalayan foothills, that constitutes Royal Chitwan National Park.

Chitwan Park sits at the confluence of three rivers, its stately sal trees full of monkeys, its brush rustling with wild peacock and deer and rhino, its waters patrolled by the world's rarest species of crocodile. On the south bank of one of those rivers is a small cluster of buildings called Tiger Tops Jungle Lodge. From there it is only a short walk to the ravine in which Bangi Bhale, on certain nights of his own choosing, feeds peacefully upon the carcass of the dead buffalo.

I was invited out to visit that ravine by InnerAsia, a travel company in San Francisco that does the American booking for Tiger Tops Lodge. From Kathmandu to the little grassy airfield at Chitwan Park is just a half-hour flight; then another two hours cross-country by elephant (a most appealing form of airport limousine) brings you to Tiger Tops. The lodge compound itself consists of a pair of thatched-roof and grass-walled buildings with rooms for about forty guests, a dining hall and a few other structures, all of these lit in the evening only by kerosene lamps. No electricity, no telephone. During monsoon season, no passable roads to the outside world. Nevertheless at this place, take my word, you are in the lap of luxury. And besides allowing wild tigers to play a heroic role in Nepal's balance-of-trade situation, the lodge has served another useful purpose: For the past ten years it has been the headquarters of Dr. Charles McDougal, one of the world's leading students of tiger behavior in the wild.

Chuck McDougal is a mild, unpretentious man with white hair and slightly stooped shoulders who has spent a large portion of his adult life crouched within grass blinds or perched in the branches of trees, in order to study the habits of *Panthera tigris*. His book, *The Face of the Tiger* (hard to find in this country but a steady seller in Kathmandu), may be the best single treatment of the subject. It is full of original insights, yet plain-spoken and judicious; it is also a loving portrait, with the sentiment left implicit. *The Face of the Tiger* was foremost among those books that had established the animal in my own mind's eye before I got anywhere near that moonlit ravine. Thanks largely to Chuck McDougal, I understood the Bengal tiger to be a majestic and complex beast—solitary by the necessities of its habitat but sociable on occasion, territorial yet peripatetic, playful, shy, patient, independent, awesomely powerful but keenly sensible. Still, I had never seen one. Not in the wild. For me the free-living Bengal tiger was just a large and beautiful idea.

Then it happened like this:

I am up in my kerosene-lit room, after dark, when a brass bell begins ringing. The alert. A tiger has killed the tethered bait. I run for the stairs, cross the compound at a trot, find the two open Land Rovers already idling, already filling with people. Climb aboard into the last open space. The Land Rovers cover a fast mile through the night, down a dirt road to a trail head. From there we start walking. Clear sky, bright crescent moon, cool air, a soft sandy trail winding uphill through sal forest. One tracker at the front, one behind, gringos like me in between. The night belongs to the tiger.

Halfway up we stop and, without a word spoken, according to the instructions we've received in advance, rip off our shoes. Leave them there. Pitter-pat onward. Twelve tourists going pitter-pat through a Nepalese jungle. It seems like some sort of Hindu religious observance. *Removing their sandals, the faithful proceed barefoot toward the sanctum.* As silently as one dozen urban people could ever be expected to do anything, we enter the blind. A small window for each of us. Eighty yards below, revealed by a spotlight that seems to bother him not at all, the big tiger named Bangi Bhale chews at his freshly killed buffalo.

After three minutes the spotlight goes off, our signal to quietly leave. Let Bangi Bhale continue his meal in peace.

I have had only a stolen glimpse. I could barely make out which end of the buffalo he was eating. I have learned nothing at all new about tiger behavior or ecology. I might have seen more in a zoo. No matter. I have come as an acolyte; and glad of it.

# SANCTUARY

## *William Faulkner and the California Condor*

The greatest and most curmudgeonly of American novelists once allowed as to how, if reincarnation were mandatory, he would prefer to come back as a buzzard. "Nothing hates him, or envies him, or wants him, or needs him," said William Faulkner. "He's never bothered or in danger. And he can eat anything." Faulkner said a lot of wild things, even sometimes when he was sober. Most of his own life was spent struggling desperately against convention and complacence and incomprehension and debt, supporting a houseful of hapless relatives, indentured as a Hollywood hack for seven years of his prime, his masterpieces already out of print when he was forty-seven years old; so maybe it's understandable, his envy for that bird he said no one envied. But if his wistful projection is set upon the greatest and most curmudgeonly of North American buzzards, *Gymnogyps californianus*, then even Faulkner, with his congenital hatred for safe opinions, was never more wildly wrong. Because the California condor, definitely, is bothered and in danger.

And the worst part is this: We don't even know by what, or how badly.

More than 10,000 years before Europeans appeared on this continent, condors were cruising the skies above desert and mountain from Texas to British Columbia. They may have been plentiful in those days, when large carnivores like the sabre-toothed cat and the dire wolf were preying on large herbivores like the mastodon and the bison, leaving scraps for scavenger birds. Some fatalistic people choose to believe in this bygone plenitude, claiming on slim evidence that the condor's decline toward extinction has been proceeding inexorably for a hundred centuries. They class the California condor as a "senile species": old and outmoded and therefore doomed. But it isn't necessarily so.

Very possibly these oddly specialized birds—dependent on a meat diet but unable to kill, graceful in the air but clumsy on landing and takeoff, nesting not in trees but in remote mountain caves—were rare even back then, during the Pleistocene epoch, when the living was easier. A man named Reuben Field, of the Lewis and Clark expedition, introduced them to the modern age on November 18, 1805, by blasting one out of the air with his rifle. Since that date, things have gone from bad to worse. In 1953 the leading scientific authority, a Berkeley ornithologist named Carl B. Koford, estimated that in all of America, all of the world, just sixty California condors remained. Today the guess is thirty.

The projection for 1995 is zero.

Throughout the past forty years the extinction of the condor has been looming closer and closer, heartbreakingly predictable, and we don't seem to be capable—by any means applied so far—of halting the trend. Now there is even some questioning, from a vociferous but small minority of environmentalists, about whether we should try. At the Condor Research Center in Ventura, a sophisticated plan has been developed for aiding the condor, a plan that would involve breeding some new birds in captivity (for eventual release to the wild) while studying others (still at large) with radio tracking. David Brower and his organization, Friends of the Earth, are vehemently opposed; the California Fish and Game Department, with legal power to prohibit this plan or sanction it, has also been dubious. One representative of the Sierra

Club has suggested that the condor, rather than being held back from the brink of extinction by elaborate scientific intervention that includes captive breeding, should instead be allowed "death with dignity" in the wild. Nearly all other concerned organizations, from National Audubon to the birding societies you've never heard of, favor intervention.

No one can say, of course, what the birds' preference on this point might be. For better or worse, just or unjust, the decision is being made by humans.

When Carl Koford was doing his study, the main concentration of nesting condors was in the canyons around Sespe Creek, a rugged outback of Ventura County less than a hundred miles from Los Angeles. From that retreat the birds ranged northward on feeding missions along the crests of the Coastal Range and the Sierras, sometimes as far as Fresno or San Jose, covering (discreetly and from far overhead) an area of roughly 50,000 square miles. With a wingspan of almost ten feet and good anatomical structure for holding those wings outspread over lengthy periods, but only modest musculature for flapping them, the condor is designed for long-distance gliding. A bird might soar off on a forty-mile sweep toward Bakersfield, watching for dead squirrels below, and soar back to its nest before sundown. This extreme mobility over a huge range, along with the condor's reclusiveness, is what makes the species so problematic to study. But Koford was fortunate in having the Sespe Creek area, dotted with nests, where he sometimes observed up to thirty condors in a single flock. In 1951 the Forest Service proclaimed 53,000 acres of this condor terrain to be the Sespe Condor Sanctuary.

But during the 1970s, while another biologist studied the condor for the U.S. Fish and Wildlife Service, no flock of more than eight birds was ever observed at the sanctuary. And the chief FWS biologist now assigned to the condor, a likable and impassioned man named Noel Snyder, has never seen, in the Sespe, a gathering of more than three. Through the past fifteen years the Sespe has survived any number of threats—a proposed dam, a proposed road, the prospect of oil and gas leasing, of mineral exploration, of continued access to hunters. Today the Sespe Sanctuary contains no oil rigs, no phosphate mines, no dams. No hunters, unless an occasional furtive poacher. No low-flying aircraft buzzing by on the way into L.A. And precious few condors.

They are dying, from whatever causes, faster than they can be born. Or they are failing to be born, for whatever reasons, as prolifically as they normally would be. Or both. Something is dragging the California condor toward extinction, some factor or combination of factors that is probably human-induced and can possibly be humanly rectified. It might be shooting or poisoning or disturbance of nesting areas or collision with power lines or hunger. But whatever it is, it can't be rectified until it is discovered. And it can't be discovered, argues Noel Snyder, using the old-fashioned, noninterventionist, observational methods. "The birds are just lost in this vast country, and you see them once in a while, but you don't study them just by seeing them fly by. You've got to come to grips with a lot more than that if you're ever going to evaluate what's going on." Even with diligent spying through the best of high-power spotting scopes, in fact, it's impossible to determine so much as a condor's sex.

Radio tracking is necessary to reveal patterns of movement, habitat requirements, frequency of breeding, survival rates, and causes of death. Such information should explain why the species is in decline. A limited program of captive breeding is also necessary because, without it, the condor population might flicker out to extinction just as the mystery is solved. But these interventionist methods certainly aren't without risk. In June of 1980, just a month after Snyder and his colleagues at the Condor Research Center received their permit for the study, one condor chick died of a heart attack while being handled.

That chick represented half of the entire known population of new-born condors for the year. There was hell to pay, and the scientists paid it: Their permit for the study was suspended. For one and a half years thereafter, the condor recovery plan remained in abeyance.

The money was there, and the scientists, and the research plan, but no California permit. (The Condor Research Center is co-sponsored by the U.S. Fish and Wildlife Service and the National Audubon Society, with financial support from federal taxpayers and the members of Audubon; the state of California, so jealous of its condors, so fastidious over the permit, contributes not a penny.) Meanwhile, time, and presumably more condors, have been lost. But now, after long negotiations, and some diplomatic compromise over the scope and methods of their study, the condor researchers have been allowed to begin again.

Maybe Noel Snyder and his co-workers will be able to discover the condor's problem in time—a parasite, or demented humans with deer rifles, or a poison used on ground squirrels, vacation homes engulfing habitat, DDT, the Los Angeles smog. Maybe they won't. Maybe time will run out. Maybe there simply can be no adequate sanctuary for such a creature, at this end of this century in this country. Maybe another chick will die, or two, or several adults, in the process of our finding that out.

It is a hard call, but aggressive and technological condor research seems to me worth the risk. Koford's observational methods didn't answer the scientific questions that need answering. Forty years of waiting and hoping hasn't helped. And the bird deserves better than this policy of passive euthanasia espoused in the name of allowing it "death with dignity." The condor is *not* a terminal victim of lymphoma. The condor's condition has yet to be diagnosed.

In December of 1950, at a certain banquet in Stockholm, William Faulkner said, "It is his privilege to help man endure by lifting his heart," and he was talking about the role of the novelist, not the role of the buzzard. But the thought applies. It is this bird's privilege to help man endure by lifting his heart. And our privilege likewise to help it, the condor, if we possibly can.

# THE
# LAST BISON

*America's Only Remnant
of Wild Buffalo*

It happened quickly. First there were 60 million, roaming the prairies and plains, blanketing whole valleys almost shoulder to shoulder for miles, the greatest abundance of any species of large mammal that modern humankind ever had the privilege to behold. And then, in 1889, there were (by one informed estimate) just 541 bison surviving throughout all the United States.

The slaughter had been conducted with prodigious efficiency and prodigious waste. Sometimes the meat was taken from a dead bison, sometimes only the hide, sometimes no more than the tongue, cut out and pickled in brine, to be sent to New York in a barrel. Sometimes not even that: People shot them from train windows to relieve the boredom of crossing Nebraska by rail, and left them rotting untouched. In the 1870s, the wildest years of the carnage, certain booking agents for the railroads went so far as to advertise outings on that basis: "Ample time will be had for a grand BUFFALO HUNT. Buffaloes are so numerous along the road that they are shot from the cars nearly every day. On our last excursion our party killed twenty buffaloes in a hunt of six hours! Round trip tickets from Leavenworth, only $10!" In Montana and the Dakotas, last refuge of the big herds, the trade in hides peaked around 1882 and then suddenly, two years later, the professional hunters were coming back from a frustrating season having seen no buffalo. None. They were gone or in hiding. Perhaps a final few desperate animals had retreated to high country, beyond the Absaroka Mountains, into Yellowstone Park. At this point among the thrill-seekers, the railroad excursionists, those idle souls back in Wichita and St. Louis and Philadelphia who collected trophies and fancied themselves "sport hunters," there was a measure of interest in that supposed distinction which would attach to the man who killed the last American bison.

But no one did. Miracle of our good fortune: No one did.

Why not? Partly because of natural human sloth: As bison grew more rare, the stalking of one became a matter of greater expense and inconvenience. Partly also because of collective good sense: Laws (belated and, at first, weak) were passed. And partly the last of the bison survived because they were not, for even an experienced and relentless hunter, so very easy to find. During that near brush with extinction at the end of the 1880s, when the species had fallen in this country to fewer than 600 individuals, and not many more in Canada, the high mountain meadows and steep woodlands of the Yellowstone plateau *did* shelter bison— probably more than 200 head, one-third of the entire national remnant.

These Yellowstone animals were not newcomers, however, not fugitives lately arrived in flight from the massacre below. They were a distinct subspecies now known as mountain bison. They had been there all along.

And they were a little different, the mountain bison, a little more cagey than their lowland relatives, perhaps more than a little better adapted to avoid terminal confrontation with man. Fossil evidence shows that they were slightly larger, on average, than plains bison (which is to say, larger than any animal on the continent), and yet from historical accounts we hear also that they were more agile and alert and wary. One observer in 1877 wrote: "These animals are by no means plentiful, and are moreover excessively shy, inhabiting the deepest, darkest defiles, or the craggy, almost precipitous, sides of mountains, inaccessible to any but the most practised mountaineers." Another writer, the park's superintendent in 1880, judged them "most keen of scent and difficult of approach of all mountain animals." The cloak of hair over their shoulders and hump was darker and finer than on plains buffalo, the alignment of horns was minutely different and, most important, the mountain bison were more hardy.

They had the evolved capability to endure those bitter and long winters in the high Yellowstone valleys—above 7,500 feet with deep snow and temperatures often below minus 25°—where a buffalo hunter, white or Indian, could too easily freeze to death in pursuit. They would face into a driving blizzard in open country and stand their ground—waiting, enduring, indomitable. They were living exempla of the word *stalwart*. They would plow snow aside with the muzzles of their massive heads to reach edible grass underneath. They would use the Firehole River and other natural geothermal features of Yellowstone as highways and oases during the worst of the winter. And in summer they climbed still higher, escaping the biting insects, grazing the sedges and grasses of subalpine meadows, venturing even onto the alpine tundra above timberline. Hannibal would have worshipped these creatures.

But despite their reclusiveness, despite their agility and power, despite the legislation that in 1872 had made Yellowstone our first national park, the mountain bison were still poached for their heads and their hides. Snowshoes and Sharps rifles made this possible, if not easy, and trophy heads were now bringing high enough prices to justify the ordeal. It was illegal but the law allowed only token penalties, and the park budget allowed only token enforcement. In 1894, after an especially flagrant poaching case was reported in the journal *Forest and Stream*, spawning further coverage in newspapers around the country and a tardy accession of public concern, Congress finally passed a law with penalties severe enough to protect the Yellowstone bison. Yet by then it was very nearly too late. Enforcement was still difficult in the Yellowstone backcountry, and by 1897 the entire park population had shrunk to less than twenty-five. These few animals were burdened with a double distinction. They were not only the last of the mountain bison. They were also the last wild bison, of any sort whatever, in all the United States.

Elsewhere the sole survivors were plains bison that had been preserved by enterprising ranchers for commercial stock-growing experiments. These private herds were dealt with like cattle: fed out on hay during hard weather, gathered periodically into corrals, the excess male calves castrated into steers. Saddled, some of them, for the amusement of their owners. Consigned to performing in rodeos. Cross-bred with domestic cattle. Doted on as nostalgic curios. And routinely slaughtered for their meat. When the century turned, there were still many buffalo in the United States, and the number increasing, but the only wild and free-living holdouts were those two dozen in Yellowstone.

And then in 1902, with well-meaning folk convinced that the little group was doomed, stock-ranching practices came also to Yellowstone. Congress put up $15,000, twenty-one plains bison were purchased from private herds in Texas and Montana, and an official "Buffalo Ranch" was established in the gorgeous Lamar Valley of the park's northeast corner. Hay was doled out, there were corrals and roundups, castrations and cullings. It became—judged on its own terms—a successful operation. Many bison were raised at the Lamar Buffalo Ranch. Only hindsight could have shown us that it was an utterly superfluous enterprise.

Superfluous because, while this ranching proceeded in the Lamar, the two dozen wild bison went their own way, to the high woodlands and the tundra in summer, to the sheltered valleys and thermal areas in winter, and survived. Left alone, given nothing but peace, they saved themselves. Endured, as they always had done, and after two decades on the brink of extinction, began again to multiply naturally.

The Buffalo Ranch is long since defunct. Its buildings now house a thriving institute for the study of Yellowstone's ecosystem. And today in Yellowstone Park, along the Lamar and the Firehole, amid the bunchgrass and sage of the Hayden Valley, across the Mirror Plateau above Specimen Ridge and at the head-waters of the Bechler River, there live about two thousand bison.

Despite some past interbreeding with—adulteration by—the old Lamar herd of coddled flatland outsiders, the Yellowstone animals represent our best and only remnant of wild bison, mountain bison, America's most imposing and resolute and dignified beast. These creatures were made for greatness. They were made to scale the spine of a continent, on tiny hooves below huge shoulders, and stand facing the driven snow. They were made to last.

# ANIMAL RIGHTS AND BEYOND

*The Search for a New Moral
Framework and a Righteous Gumbo*

Do non-human animals have rights? Should we humans feel morally bound to exercise consideration for the lives and well-being of individual members of other animal species? If so, how much consideration, and by what logic? Is it permissible to torture and kill? Is it permissible to kill cleanly, without prolonged pain? To abuse or exploit without killing? For a moment, don't think about whales or wolves or the California condor; don't think about the cat or the golden retriever with whom you share your house. Think about chickens. Think about laboratory monkeys and then think about lab rats and then also think about lab frogs. Think about scallops. Think about mosquitoes.

It's a Gordian question, by my lights, but one not very well suited to Alexandrian answers. Some people would disagree, judging the matter simply enough settled, one way or the other. *Of course they have rights. Of course they don't.* I say beware any such snappy, steel-trap thinking. Some folk would even—this late in the evolution of human sensibility—call it a frivolous question, a time-filling diversion for emotional hemophiliacs and cranks. *Women's rights, gay rights, now for Christ sake they want ANIMAL rights.* Notwithstanding the ridicule, the strong biases toward each side, it is certainly a serious philosophical issue, important and tricky, with almost endless implications for the way we humans live and should live on this planet.

Philosophers of earlier ages, if they touched the subject at all, were likely to be dismissive. Thomas Aquinas announced emphatically that animals "are intended for man's use in the natural order. Hence it is no wrong for man to make use of them, either by killing or in any other way whatever." Descartes held that animals are merely machines. As late as 1901, a moral logician named Joseph Rickaby (who happened to be a Jesuit, but don't necessarily hold that against him) declared: "Brute beasts, not having understanding and therefore not being persons, cannot have any rights. The conclusion is clear." Maybe not quite so clear. Recently, just during the past decade, professional academic philosophers have at last begun to address the matter more open-mindedly.

Two thinkers in particular have been influential: an Australian named Peter Singer, an American named Tom Regan. In 1975 Singer published a book titled *Animal Liberation*, which stirred up the debate among his colleagues and is still treated as a landmark. Eight years later Tom Regan published *The Case for Animal Rights*, a more thorough and ponderous opus that stands now as a sort of companion piece to the Singer book. In between there came a number of other discussions of animal rights—including a collection of essays edited jointly by Singer and Regan. Despite the one-time collaboration, Peter Singer and Tom Regan represent two distinct schools of thought: They reach similar conclusions about the obligations of humans to other animals, but the moral logic is very different, and possibly also the implications. Both men have produced some formidable work and both, to my simple mind, show some shocking limitations of vision.

I've spent the past week amid these books, Singer's and Regan's and the rest. It has been an edifying experience, and now I'm more puzzled than ever. I keep thinking about monkeys and frogs and mosquitoes and—sorry, but I'm quite serious—carrots.

*   *   *

Peter Singer's view is grounded upon the work of Jeremy Bentham, that eighteenth-century British philosopher generally known as the founder of utilitarianism. "The greatest good for the greatest number" is a familiar cartoon version of what, according to Bentham, should be achieved by the ethical ordering of society and behavior. A more precise summary is offered by Singer: "In other words, the interests of every being affected by an action are to be taken into account and given the same weight as the like interests of any other being." If this much is granted, the crucial next point is deciding what things constitute *interests* and who or what qualifies as a *being*. Evidently Bentham did not have just humans in mind. Back in 1789, optimistically and perhaps presciently, he wrote: "The day *may* come when the rest of the animal creation may acquire those rights which never could have been withholden from them but by the hand of tyranny." Most philosophers of his day were inclined (as most in our day are still inclined) to extend moral coverage only to humans, because only humans (supposedly) are rational and communicative. Jeremy Bentham took exception: "The question is not, Can they *reason?* nor, Can they *talk?* but, Can they *suffer?*" On this crucial point, Peter Singer follows Bentham.

The capacity to suffer, says Singer, is what separates a being with legitimate interests from an entity without interests. A stone has no interests that must be respected, because it cannot suffer. A mouse can suffer; therefore it has interests and those interests must be weighed in the moral balance. Fine, that much seems simple enough. Certain people of sophistic or Skinnerian bent would argue that there is no proof a mouse can in fact suffer, that it's merely an anthropomorphic assumption; but since each of us has no proof that *anyone* else actually suffers besides ourselves, we are willing, most of us, to grant the assumption. More problematic is that very large gray area between stones and mice.

Peter Singer declares: "If a being suffers, there can be no moral justification for disregarding that suffering, or for refusing to count it equally with the like suffering of any other being. But the converse of this is also true. If a being is not capable of suffering, or of enjoyment, there is nothing to take into account." Where is the boundary? Where falls the line between creatures who suffer and those that are incapable? Singer's cold philosophic eye travels across the pageant of living species—chickens suffer, mice suffer, fish suffer, um, lobsters most likely suffer, *look alive, you other creatures!*—and his damning stare lands on the oyster.

No I'm not making this up. The oyster, by Singer's best guess, doesn't suffer. Its nervous system lacks the requisite complexity. Therefore, while lobsters and crawfish and shrimp possess inviolable moral status, the oyster has none. It is a difficult judgment, Singer admits, by no means an infallible one, but "somewhere between a shrimp and an oyster seems as good a place to draw the line as any, and better than most."

Moral philosophy, no one denies, is an imperfect science.

Tom Regan takes exception with Singer on two important points. First, he disavows the utilitarian framework, with its logic that abuse or killing of animals by humans is wrong because it yields a net *overall* decrease in welfare, among all beings who qualify for moral status. No, argues Regan, that logic is false and pernicious. The abuse or killing is wrong in its *essence*—however the balance comes out on overall welfare—because it violates the rights of those individual animals. Individual rights, in other words, take precedence over the maximizing of the common good. Second, in Regan's opinion the capacity to suffer is not what marks the elect. Mere suffering is not sufficient. Instead he posits the concept of *inherent value*, a complex and magica quality possessed by some living creatures but not others.

A large portion of Regan's book is devoted to arguing toward this concept. He is more uncompromisingly protective of certain creatures—those with rights—than Singer, but he is also more selective; the hull of his ark is sturdier, but the gangplank is narrower. According to Regan, individual beings possess inherent value (and therefore inviolable rights) if they "are able to perceive and remember; if they have beliefs, desires, and preferences; if they are able to act intentionally in pursuit of their desires or goals; if they are sentient and have an emotional life; if they have a sense of the future, including a sense of their own future; if they have a psychophysical identity over time; and if they have an individual experiential welfare that is logically independent of their utility for, and the interests of, others." So Tom Regan is not handing rights around profligately, to every cute little beast that crawls over his foot. In fact we all probably know a few humans who, at least on a bad night, might have trouble meeting those standards. But how would Regan himself apply them? Where does he see the line falling? Who qualifies for inherent value, and what doesn't?

Like Singer, Regan has thought this point through. Based on his grasp of biology and ethology, he is willing to grant rights to "mentally normal mammals of a year or more."

Also like Singer, he admits that the judgment is not infallible: "Because we are uncertain where the boundaries of consciousness lie, it is not unreasonable to advocate a policy that bespeaks moral caution." So chickens and frogs should be given the benefit of the doubt, as should all other animals that bear a certain degree of anatomical and physiological resemblance to us mentally normal mammals.

But Regan does not specify just what degree.

The books by Singer and Regan leave me with two very separate reactions. The first combines admiration and gratitude. These men are applying the methods of systematic philosophy to an important and much-neglected question. Furthermore, they don't content themselves with just understanding and describing a pattern of gross injustice; they also emphatically say *Let's stop it!* They are fighting a good fight. Peter Singer's book in particular has focused attention on the outrageous practices that are routine in American factory farms, in "psychological" experimentation, in research on the toxicity of cosmetics. Do you know how chickens are dealt with on the large poultry operations? How veal is

produced? How the udders of dairy cows are kept flowing? Do
you know the sorts of ingenious but pointless torment that thousands of monkeys and millions of rats endure, each year, to fill
the time and the dissertations of uninspired graduate students?
If you don't, by all means read Singer's *Animal Liberation*.

The second reaction is negative. Peter Singer and Tom Regan,
it seems to me, share a breathtaking smugness and myopia not
too dissimilar to the brand they so forcefully condemn. Theirs is
a righteous and vigorous smugness, not a passive and unreflective
one. But still.

Singer inveighs against a sin he labels *speciesism*—discrimination against certain creatures based solely upon the species to
which they belong. Regan uses a slightly less confused and less
clumsy phrase, *human chauvinism*, to indicate roughly the same
thing. Both of them arrive (supposedly by sheer logic) at the
position that vegetarianism is morally obligatory: To kill and eat
a "higher" animal represents absolute violation of one being's
rights; to kill and eat a plant evidently violates nothing at all.
Both Singer and Regan claim to disparage the notion—pervasive
in Western philosophy since Protagoras—that "Man is the measure of all things." Both argue elaborately against anthropocen-
trism, while creating new moral frameworks that are also decidedly
anthropocentric. Make no mistake: Man is still the measure, for
Singer and Regan. The test for inherent value has changed only
slightly. Instead of asking *Is the creature a human?*, they simply
ask *How similar to human is similar enough?*

Peter Singer explains that shrimp deserve brotherly treatment
but oysters, so different from us, are fair game for the gumbo.
In Tom Regan's vocabulary, the redwood tree is an "inanimate
natural object," sharing that category with clouds and rocks. But
some simple minds would say: Life is life.

# THE
# BIG GOODBYE

*Who Will Survive
the Late Quaternary Extinction?*

There are extinctions, and then again there are Extinctions.

Inevitably every once in a while a single species passes quietly into oblivion. At other and much rarer times large groups of species—entire genera and families of animals and plants, entire civilizations of interrelated organisms—disappear suddenly in a great catastrophic wipeout. During the Permian Extinction, for instance, roughly 225 million years ago, half of all the families of marine creatures (which were then the predominant form of life) died away in a brief few million years. No one knows why, and the question is still debated, but most likely the cause was habitat loss, when the rich oceanic shelfs were left high and dry by falling sea levels. The Cretaceous Extinction, 65 million years ago, was equally drastic and even more puzzling: After more than 100 million years of unrivaled success, the dinosaurs rather abruptly disappeared, as did the various flying and fishlike reptiles, and many more groups of marine invertebrates. Again there is no proven explanation but the suggested causes include global temperature change, reversal of the polar magnetic field, and the impact of a hypothetical asteroid six miles across which raised such an atmospheric dust cloud that no sunlight could penetrate and no green plants could grow for ten years. Finally and most dramatically, the Late Quaternary Extinction, during which more than a million species of living things perished within just a century. This quickest of all mass extinctions occurred (according to the local time system) in the span 1914–2014 A.D. The main cause was once again habitat loss, and the agent of that loss was the killer-primate *Homo sapiens*, now itself extinct. *Sapiens* unaccountably violated the first rule of a successful parasite: moderation. *Sapiens* was suicidally rapacious.

That's the way it will look to some being on the planet Tralfamadore with an idle interest in the paleontology of Earth. Life has existed on this mudball for about 3½ billion years, and we are just now in the midst of what looks to shape up as the third great mass extinction of species. This episode threatens to be larger in consequence than the Permian and the Cretaceous and the other major die-offs put together: One-fifth of all forms of earthly organism could be gone within thirty years.

After that, things would get ugly for the survivors. Global climatic conditions would change, with accelerated buildup of carbon dioxide in the atmosphere, disruption of wind currents, cycles of vastly increasing erosion despite decreasing rainfall, breakdown of natural processes for the purification of fresh water, warmer average temperatures, the eventual failure of domestic food crops—and that would only be the beginning. We have all heard about snail darters and whooping cranes until our eyes glaze over, but what in fact is at issue here is the overall biological stability of a world. The Late Quaternary Extinction wants *you*.

In a broad sense the LQE began about 400 years ago, with the European age of empires, when humankind reached a stage smart enough to sail all over the planet and still stupid enough to kill much of what we found when we got there. Dutch settlers arrived on the island of Mauritius in 1598 and the dodo was extinct by 1681. On Bering Island off Alaska, the last Steller's sea cow was killed by a party of Russian sealers in 1768. Icelandic hunters

killed a lonely pair of great auk in June of 1844, and no great auk
has been seen alive since. But these were just the preliminaries.
In a stricter sense the start of the Late Quaternary Extinction can
be set, with precision that is artificial but emblematic, as September 1, 1914. That day the last passenger pigeon, name of
Martha, died in the Cincinnati Zoo.

Martha is significant because her species—despite incredibly
intense hunting pressure against them—succumbed chiefly to loss
of habitat. The passenger pigeon, which had once been perhaps
the most numerous bird on earth, needed huge, continuous areas
of oak and beechnut forest for its gregarious patterns of feeding
and nesting. With the great hardwood forests east of the Mississippi cut back to small pockets, the passenger pigeon had no more
chance of surviving than, literally, a fish out of water. And at
this end of the century the same thing is happening, say eminent
biologists like Thomas Lovejoy and Norman Myers, to hundreds
of species—*poof:* gone forever—each year.

Eventual extinction is as natural for every species as eventual
death is for every individual creature. What matters for biological
stability are (1) patterns of extinction, and (2) rate of extinction.
As long as extinctions occur no more rapidly than new species
arise, and are not so clustered in particular areas as to destroy the
conditions from which new species *can* arise, then ecosystems
remain stable and healthy. While the Cretaceous Extinction was
in full swing, paleontologists estimate, the rate of disappearance
was one species every thousand years. Between 1600 and 1900
A.D., with our improved capabilities for travel and hunting, mankind eliminated roughly seventy-five species of known mammals
and birds. Since 1900 we have killed off another seventy-five
species of conspicuous animal—just less than one per year. For
a single new species of bird to diverge from another species probably takes at least 10,000 years.

A bad balance, but growing still worse: Norman Myers, having studied the problem for years from his base in Nairobi, figures
we might say goodbye to *one million* further species by the year

2000. That amounts to about 100 species driven extinct every day.

Numbers, yes. Boggling, dulling numbers. Finally it doesn't sound real. *What* are *all these vanishing species? Where are they? And what's killing them?*

But it is real. They are, for the most part, inconspicuous but ecologically crucial organisms: plants, insects, fungi, crustaceans, mites, nematode worms. They inhabit those ecological zones that are richest in living diversity but have been least investigated by man: estuaries, shallow oceanic shelfs, coral reefs, and in particular, tropical rainforests. They are being extinguished, like the passenger pigeon, by human activities that alter their habitats.

We are poisoning the estuaries with our industrial and municipal wastes, we are drilling for oil and spilling it on the ocean shelfs; but most egregious and most critical is the destruction of tropical rainforests. Rainforests comprise only 6 percent of the Earth's land surface, yet may hold as many as half the Earth's total number of species, and two-thirds of all species of plant. The rainforests of Central America and the Amazon are today being mown down for pulpwood, and to graze cattle on the cleared land so that American hamburger chains can buy cheap beef. Rainforests of the Philippines, Indonesia and other parts of tropical Asia are being lumbered, to fill the demand for plywood and exotic hardwoods in more affluent countries. In West Africa the forests are falling chiefly to slash-and-burn agricultural methods by starving peasants who can't feed their growing families off small permanent fields, partly because world oil demand has priced them out of any chance for petroleum-based fertilizer. Altogether the planet may be losing 3,000 acres of rainforest—and four irreplaceable living species—every hour. We are gaining rosewood and mahogany trinkets, profligate use of personal autos, and the Big Mac.

There is complicity involved here, more than a share for us all. Norman Myers says that "the main problem for declining wildlife is not the person with conscious intent to exploit or kill:

it is the citizen who, by virtue of his consumerist lifestyle, stimulates economic processes that lead to disruption of natural environments."ments."

All of this is tenaciously intertangled—the guilt, the patterns of demand, the good selfish considerations that should dictate species preservation—as intertangled as the life cycles of the species themselves.

Plants, for instance. The educated guess is that each species of plant supports ten to thirty species of dependent animal. Eliminate just one species of insect and you may have destroyed the sole specific pollinator for a flowering plant; when that plant consequently vanishes, so may another twenty-nine species of insects that rely on it for food; each of those twenty-nine species might be an important parasite upon still another species of insect, a pest, which when left uncontrolled by parasitism will destroy further whole populations of trees, which themselves had been important because. . . .

And so on, into the endless reticulation, the endless fragile chain of interdependence that is a tropical ecosystem. Of course it is possible that the most dire projections will not become reality, that the trend will change, that mankind at the last minute will show unexpected forbearance—as economist Julian Simon and other bullish anthropocentrists are fond of predicting. Possible, yes. It is *possible*, for that matter, that with a hundred years of trying genetic engineers at the General Electric laboratories might find a way to re-invent the passenger pigeon. (If so, they would probably patent it.) Possible but not likely. Neither of those cheerful miracles can sensibly be counted on.

What seems all *too* likely is that the present trend will continue; that mankind will have cut down and bulldozed away most of the world's rainforests before the year 2000. If so, by direct action alone we will have thereby exterminated perhaps 150,000 species of plants—with indirect consequences, among other creatures and for the biosphere as a whole, that would be geometrically larger. Maybe 900,000 species of insect lost; and the 291 species of tropical bird already known to be threatened; and the fewer than fifty remaining Sumatran rhino; and the nine surviving representatives of the Mauritius kestrel. Enough numbers.

No, one final number: 300. Just that many years have passed since 1681. In this tricentennial year of the extinction of *Raphus cucullatus*, the giant flightless Mauritian pigeon, it is worth remembering that *Homo sapiens* too could become part of the Late Quaternary Extinction, engineering ourselves a place among the next group of species bidding this planet the Dodo's Farewell.

# ELOQUENT PRACTICES, NATURAL ACTS

# LOVE'S MARTYRS

*The Metaphysical Poetry
of Semelparity*

Like Woody Allen, the English poet John Donne was, in his younger days, obsessed with love and death. Throughout Donne's early work those two motifs recur again and again, linked so closely together that they come to seem logically inseparable, two sides of a macabre equation that evoke each other almost interchangeably. Love is a manner of dying, Donne suggests; and vice versa. For instance: "I cannot say I lov'd, for who can say/ Hee was kill'd yesterday?" Another poem contains the tender sentiment: "Since thou and I sigh one anothers breath,/ Who e'r sighes most, is cruellest, and hastes the others death." In still another, a man speaking from his own grave declares himself "love's martyr," dead of an excess of passion. During the sixteenth century in England, this oxymoronic linkage of the two concepts—love and death—was enough to get an ardent young poet eventually categorized as one of the "metaphysical" school. Nowadays it would make him a theoretical population ecologist.

The notion with which John Donne was flirting, in all that love-and-death poetry, is studied today as a curious corner of evolutionary ecology, under the label *semelparity*. The word semelparity, of course, is just another bit of that formal jargon

149

scientists take cruel joy in inventing. The same thing is more casually known (with a lewd nod from the ecologists to their cosmologist colleagues) as Big-Bang reproduction.

Semelparity: An animal or a plant waits a very long time to breed only once, does so with suicidal strenuosity, and then promptly dies. The act of sexual procreation itself proves to be ecstatically fatal, fatally ecstatic. And the rest of us are left merely to say: Wow.

As a strategy for self-perpetuation, semelparity is still only marginally understood. But the list of known semelparous creatures is intriguingly diverse. Bamboo do it. A group of hardy desert plants called the agave do it. Pacific salmon do it. The question is why. What can these three organisms, apparently so dissimilar, have in common? Why should all three—living in drastically different environments with drastically different life histories—be similarly committed to dying for one taste of love?

The answer, according to cautious speculation by some ecologists, may be as simple as a few symbols. Natural selection, these researchers say, tends to maximize for each age i the sum $B_i + p_i(v_{i+1}/v_0)$. Now if you are like me, any such clot of algebraic nonsense immediately causes alarm bells to ring in your head, sprinkler systems to begin dousing your overheated brain, and your eyes to slide straight off the page like a cheap ballpoint skidding across oilpaper. But wait. The idea entombed in that ugly cryptogram happens to be rather interesting and, just possibly, it can also be said in English.

First, a few concrete facts.

Five different species of salmon migrate regularly from the Pacific Ocean into the rivers of western North America. They are headed upstream to spawn, and in each individual case the journey is a return trip back to that same freshwater tributary where the adult salmon began its own life. Some of these fish (the chinook of the Yukon River, for instance) will travel as much as 2,000 miles, climbing through rapids, making ten-foot leaps to clear the cascades, dodging predators, fighting constantly up-current at an unflagging pace of perhaps fifty miles a day. The effort and determination involved are prodigious, but it is a one-way trip. Soon after having spawned, every female and every male of these five species is dead. Decomposing corpses pile up in the eddies, turning clear mountain water funky with rotting flesh. Among those lucky fish that have completed the trip, won a mate, found genetic fulfillment in the gentle current above a gravel spawning nest, there are no survivors.

Certain types of bamboo make a long journey to breed also, but the distance they cover is measured in time. The common Chinese species *P. bambusoides*, for instance, has a regular life cycle of 120 years. Historical sources back into the tenth century record its episodes of massive synchronous flowering. Each time around, a dense grove of *P. bambusoides* begin life together as new sprouts; for 120 years they merely grow taller and sturdier, putting out leaves and branches, storing away energy, adding clones of them-selves to the population by a nonsexual process of budding; then after the appointed six score years they suddenly, simultaneously, produce an awesome profusion of flowers. The blossoms fertilize each other by wind. Seeds fall like heavy hail, coating the ground, ankle deep to a man. And the 120-year-old progenitors all im-mediately die.

The average agave, meanwhile, is content through long years of growth and celibacy to resemble an artichoke peeking out be-tween desert rocks. Agaves are succulents, related to lilies but preferring a hard life on the dry hillsides of the Sonoran desert. They unfold their leaves in a spiral rosette, each leaf tipped with a sharp spine for protection against browsers, and they range by species from the size of a small porcupine to the size of a 300-pound octopus. Like the bamboo, they can set off clones of them-selves by budding, but their primary mode of reproduction is sexual. Years go by uneventfully, an agave grows large and vig-orous, until with startling fanatical abandon, one season, it pro-duces a towering flower stalk. This great inflorescence will be notable not for the beauty or fragrance of its blossoms, but for its sheer height. From a big agave, in just three or four months, the stalk might shoot up thirty feet, sturdy and straight as a young lodgepole pine. It's a flower you'd want a chainsaw to cut. Pollination is done by insects and bats; seeds drop. And down below, shriveled and brown, the agave itself is already dying, as though stabbed through the heart with its own flower stalk.

Semelparity at work in the moist tropics of Asia, on the driest slopes of Baja, at the headwaters of the Yukon River. A curious person naturally wants to know why, in each case, a single act of sex should prove deadly. But even more tantalizing, it seems to me, is the first derivative of that question: Is there any *one* *answer* that can explain all three cases?

A man named William M. Schaffer says yes, and offers us this:

$$\sum_i [B_i + p_i(v_{i+1}/v_0)] = k$$

Hold the sprinklers, hold the alarm. What he is talking about is simply a delicate balance between love and death.

Dr. Schaffer is a respected theoretical ecologist at the University of Arizona. In a half dozen papers published over the past decade, alone and with various co-authors, he has proposed a theory of how evolution tends to produce optimal life-history strategies for different animals and plants. This theory entails enough mathematical ree-bop-dee-bop to make the average human swoon. Descriptive numbers for a particular species can be inserted into Schaffer's equations, a lever is then pulled, a crank is turned, and the theory will posit how the creature should act if it wants to survive the Darwinian struggle. That's the sort of thing theoretical ecologists do. Semelparity is quite useful in such theorizing, because it involves the love-and-death balance taken to one logical extreme.

Dr. Schaffer calls that balance the *trade-off function*. The particular trade-off at issue is between present and future, the immediate prospect of producing offspring versus the chance of surviving to produce other broods later. An underlying premise here is that an animal or plant has, at each stage in its life, only a limited total amount of available energy that it can spend on the business of living. That limited energy must be shared out among three fundamental categories of effort—routine metabolism, growth, and reproduction—and an optimal life history is one that balances the shares most efficiently to produce evolutionary success. Evolutionary success, of course, is measured on one simple scale: How many offspring does the creature leave behind? All this is quite basic.

Schaffer's trade-off equation merely codifies that crucial balance between present and future, between short-term concerns and a longer view—that is, between effort devoted to immediate self-preservation and effort devoted to parenthood. His $B_i$ represents the reproductive potential of a given creature right now. The factor $p_i$ stands for the probability (or improbability) that a creature which has bred once will survive to breed again later. The parenthesis $(v_{i+1}/v_0)$ will be the remaining reproductive potential which an experienced parent can expect still to possess at that hypothetical later time. Balance all those considerations against one another, with just the proper commitment of energy (at each age) to routine metabolism, to growth, to reproduction, and the result is high evolutionary success. This is the burden of $\sum_i [B_i + p_i(v_{i+1}/v_0)] = k$.

Now let's try it in English: If a female chinook salmon, having swum 2,000 miles up the Yukon River, having climbed rapids and dodged otters and leapt cascades, having beaten her fins to tatters digging a gravel nest and fought off other salmon to keep the spot inviolate—if this poor haggard creature has virtually *no chance* of surviving to accomplish the same entire feat again, then she will be *required* by the forces of natural selection to sacrifice herself totally, in one great suicidal effort of unstinting motherhood from which she cannot possibly recover. She will lay about 5,000 eggs. And then she will promptly croak.

Likewise for those Sonoran agave, for those Chinese bamboo: The terms of the trade-off are the same, the result is the same; only the numbers and the facts are different. Bamboo seem to sacrifice themselves for the sake of *predator satiation*—producing so many seeds that, after all the rats and jungle fowl of China have eaten their fill, a few seeds will still be left to germinate. The agave compete with each other to produce taller and yet taller flowers, apparently because their pollinators deign to visit only the tallest. The evolutionary consequence in each case, as with salmon, is semelparity.

But Dr. Schaffer's neat mathematical model is not without gaps, not without weaknesses. What about the *Atlantic* salmon, for instance, which faces an almost identical set of circumstances yet does *not* resort to semelparity? It can be argued that Schaffer's equations constitute an unduly abstract version of reality. To some observers, such airy theorizing has little more connection to the untidy actualities of ecological fieldwork than it does to, say, metaphysical poetry.

At Christmas of the year 1600, John Donne became secretly married to a girl for whom he felt irresistible passion. It was a dangerous step and he knew that. Donne was twenty-seven years old, son of an ironmonger; she was sixteen, daughter of a powerful noble. The young poet was dismissed from his job. Briefly he went to prison. For years afterward, with a wife and children, he was blackballed by his in-laws from earning a living. Around that time he wrote:

> *Love with excesse of heat, more yong than old,*
> *Death kills with too much cold. . .*
> *Once I lov'd and dy'd; and am now become*
> *Mine Epitaph and Tombe.*
> *Here dead men speake their last, and so do I;*
> *Love-slaine, loe, here I lye.*

It is highly unlikely that John Donne ever set eyes upon a Pacific salmon, or a Sonoran agave, or even a transplanted grove of *P. bambusoides* in some London botanical garden. But we can safely assume that, nevertheless, he would have understood.

# LIVING
# WATER

The human body consists mainly of water. So does a cucumber. So does the surface of the Earth. An odorless, tasteless liquid, say the scientific encyclopedias, abundant and widely distributed throughout the planet; built with simple elegance from two hydrogen atoms and an oxygen; essential in plant and animal nutrition; constituent of the crystals of many minerals; blue when encountered in thick layers: water. You can't get away from the stuff, and if you could, you wouldn't want to. It is an exaggeration to say that "Water is life," an exaggeration of the ubiquitousness, and perhaps the importance, certainly the durability, of life. Water came first, necessarily. Without life, there would still be water. Without water, no life.

The first primitive form of life was probably born about 3½ billion years ago, when a bolt of lightning striking the primordial sea water delivered an energy jolt that spurred dissolved amino acids to cluster into self-interested coalitions of protein. The protein clusters began to grow into droplets, to pull water molecules up around themselves like cell walls, to compete with each other for the recruitment of further dissolved molecules—and the process of organic evolution was under way. But by that time it had already been raining on Earth, the seas had been rising and falling, the rivers had been flowing, for over a billion years.

Of all the types of hydrological features into which this planet's water has arranged itself—oceans, polar ice, glaciers, groundwater, fresh and salt lakes, soil moisture, atmospheric vapor— the least imposing, judged by volume, are rivers. Only one millionth part of our total water, as it turns out, is at this moment moving within the banks of rivers. Don't be fooled: They are important, in the history of life, in the history of man, far beyond their size. Rivers have served as crucibles of evolution, pathways of colonization, sources of power, and inspiration, and topsoil. They not only provide, they deliver. Glaciers come and go, lakes fill up and disappear, but rivers continue to follow their chosen routes with exceptional permanence, changing character gradually, maturing, cutting tight canyons and then wide valleys, shifting about restlessly within their domains, always wandering, always leaving, never gone; offering to their living inhabitants that one crucial element that survival by evolution demands: time. Just recently—say within the last 7,000 years—the great early flowerings of human culture appeared, surely by no accident, in the Nile Valley, along the Indus, between the Tigris and the Euphrates, on the banks of the Hwang Ho. The sea is where we came from. Rivers are how we got here.

Every river on Earth is a symphony of elaborate and ordered entanglements—physical factors entangled with chemical factors entangled with biological ones—and as sure as the *Jupiter* symphony is not the *Pastorale*, every river is different from all others. But one thing remains common by definition, and represents the central, dominating fact to be coped with by the creatures that make their lives in rivers: The water never stops moving. The pressure of current never relents. The river, as Roderick Haig-Brown has said, never sleeps. To the plants and animals involved, the current brings food, oxygen and all the other nutrients they need for survival; it also, that very same current, threatens every moment to sweep them downstream into drastically different habitats where they will quickly die. The current giveth, the current taketh away, the current has the power of a god. Because of this harsh reality, a river is perhaps the most challenging of ecosystems. It is also one of the most complicated and—thanks to that complexity—one of the most rich.

Current velocity is a variable factor, interdependent with a number of other physical variables, including source of the water, slope of the streambed, depth of the channel, width of the channel and total volume of the water being moved. When one of those factors changes, the others make compensatory adjustments, and the path and shape of a river reflect their compromise. For instance, if the source of a river is snow-melt from high mountainsides, the volume of discharge generally rises to a torrential peak in late spring or early summer, carrying sediments and pushing boulders, back-cutting by erosion to flatten the slope of the streambed, dismantling riffles and rebuilding them, widening and deepening the channel on the outside of the bends, sometimes gouging entirely new channels, short cuts, that may become permanent when the flooding subsides. If the source is a glacier, the pulse of high water comes later, in midsummer, and is less extreme, simply because ice melts more slowly than snow. If the source is a spring, there is no annual flood, little erosion, greater permanence to the shape of the channel. And if the source is groundwater seepage, the river is actually lowered in springtime, when riverside trees leaf out and begin competing to suck away moisture. Most rivers, of course, have a combination of sources and a mixture of these flow regimens. The whole system of discharge and channel adjustment tends, as the scientists say, toward conservative dynamic equilibrium. In other words: The water looks for the laziest way of getting itself fastest to the sea.

In times of normal flow, a streambed composed of rocks and large boulders retards current velocity, holding the water back by friction, while a sandy bottom offers little resistance. A wide and deep channel also entails less friction—which is why broad lower stretches of a river like the Mississippi, despite the gentle slope, despite the sluggish appearance, despite what we expect, may well be flowing faster than many steep mountain rivers.

Whatever its size or shape, a river with current velocity of four to six feet per second is moving quite swiftly. At eight feet per second, it is roaring. Very few creatures, even among those adapted to river habitats, can survive against such a constant force. So they have found and invented ways, not to live in the current, but to live out of it. The most important of those little secrets is called the boundary layer.

Current speed at any point in a channel is inversely related to depth, with the surface water in midstream moving fastest, other water moving slower beneath the surface and nearer the banks, and the deep water moving slower still; but a point is reached, about one to three millimeters above the streambed, where the water is not moving at all. Brought to a full stop by friction, this boundary layer of still water allows a huge variety of small animals, mainly insects, to live and flourish along stream bottoms in even the stretches of heaviest current. It also explains why most river insects are shaped more like alligators than like giraffes: They are adapted for life in the boundary layer, where it is a mortal necessity to keep your head down.

Also necessary to support life on the stream bottom are a variety of dissolved gases and solids, especially oxygen, carbon dioxide, calcium, potassium, nitrate, phosphate, and silica. Unless oxygen is present to the level of four or five parts per million, animals such as stoneflies and trout, with great respiratory needs, will suffocate. Under conditions of high temperature, the same animals may require twice as much oxygen—yet as the water temperature rises above 70°F., oxygen will be forced out of solution and lost from the river. It is stirred back in by any turbulent movement, over riffles, down cascades, and replaced during daylight by the photosynthetic activity of green plants. At night the concentration tends to fall again, with photosynthesis shut down for lack of light, and both animals and plants using oxygen for respiration. The decomposition of dead plant tissue, a massive continuing enterprise, conducted in rivers by bacteria and fungi, also consumes oxygen. Rivers are constantly depleting their oxygen, and constantly—unless they are dammed—replacing it. A river must tumble, must thrash, must dance along freely, or it goes blue in the face.

Meanwhile, the plants breathe in the carbon dioxide that the animals have exhaled. Thick layers of moss spread like green terry cloth over rocks in the shady areas where carbon dioxide is plentiful, and provide habitat for tiny species of insects, crustaceans, and fleas. Dissolved carbon dioxide also nourishes algae (which serve as fodder for many insects), and allows calcium to remain in solution as calcium bicarbonate, an important mineral salt. This available calcium salt balances the acidity of the river's water, encourages algal growth, helps fish to breathe more efficiently and provides snails, mussels, and crayfish with building material for their shells. It is precious useful stuff, calcium bicarbonate.

Potassium, nitrate, and phosphate are all essential plant nutrients, natural fertilizers, and even silica evidently spurs the growth of diatoms, those microscopic algae that coat stream boulders in a smooth slime and turn wading fishermen into riotously comic spectacles. Nitrate and phosphate enter a river mainly through rainfall, runoff from the land surface and bank erosion. Nitrate also arrives, like a sprinkling of compost to help the river plants burgeon, in the dead foliage dropped on the water by streamside trees—particularly a common species of alder, which concentrates in its leaves exceptionally high levels of nitrate, and vastly enriches the food cycle of any river into which those leaves fall.

Fallen leaves, in fact, are the single chief source of fuel for the river ecosystem. The chain of river life begins—in an important sense, the sense of energy transfer and the construction of living matter—in autumn, at just the time when life on land is closing down for the winter hiatus. Yellowed alder leaves, aspen and willow and cottonwood, drift onto the water and float for a while, then sink, to become wedged between rocks; bacteria and fungi climb aboard and begin feasting, digesting, causing decay; the leaves crumble to a fine mulch that sails away on the current. Downstream, the larvae of caddisflies, blackflies and mayflies scoop in the mulch with all manner of ingenious nets and filters, and devour it like a chef's salad, bacteria and fungi included.

Where the mixture falls to rest in dead water, where half-decayed fragments catch in crannies, stonefly nymphs waddle up to browse, hungry crayfish appear, and those delicate shrimplike scudders, the amphipods. Eaten once, passed once through a gut, the same stuff is taken again further downstream by other insects and shellfish, passed again through a gut, and still again after that, until all food value has been extracted. It's a system that brooks no waste. This nutritious vegetable bounty is called detritus: the granola of rivers.

Other species of mayfly, stonefly and caddisfly, as well as midge larvae and snails, satisfy themselves grazing algae off rocks. The common blackfly larva, bizarre of design and flexible of habit, hangs backward from hooked feet with its head swinging downstream, straining the current with bristly mustaches for detritus and floating diatoms, and then occasionally, for variety, bends over to scrape its foothold clear of algae. A few stoneflies and caddisflies also raid the moss gardens. All of these pacific invertebrates—grazing and cropping and savoring their tangy mulch—are the primary consumers in the river ecosystem, the creatures responsible for turning vegetables into meat. After them, in hot pursuit, come their natural enemies, the small carnivores. Most of these, too, are insects.

The dragonfly nymph is a formidable hunter, with a wildly improbable lower lip that springs out under hydraulic pressure, snags a tiny victim in its hooked teeth, and then, on retrieve, slaps the food straight back into the dragonfly's mouth, brisk and indelicate as a chimp stealing peaches. Making it even more deadly, the nymph has a pair of large compound eyes that achieve their binocular focus at precisely the point before its nose where the lip structure reaches full extension: Any morsel seen with both eyes is a morsel perfectly targeted. One type of caddisfly larva builds a conical silken net facing open to the current, then lurks at the narrow end and, when a smaller animal is swept in, rushes out like a spider to pounce on it. Large nymphs from a branch of the stonefly clan are also estimable predators, as are both the larvae and adults of some water beetles, and even a few species of mayfly. On their best days, these secondary consumers rule as lords of the stream-bottom jungle; one bad day, one mistake, one loss of footing, and they are in the belly of a trout.

Thousands of bad days for millions of cold-water insects, and the result is what we often call, with some narrowness of vision, a good trout stream. But a good trout stream must first be an excellent insect stream, a superior haven for algae and fungi and bacteria, a prime dumping ground for dead leaves, a surpassing reservoir of oxygen and calcium. It will then also, and thereby, be a good osprey stream, a favorite among otters, a salvation to dippers and kingfishers and bank swallows and heron, mergansers and Canada geese and water shrews, mink and muskrat and beaver. Not to mention the occasional grizzly bear. And who knows but that, sometime, a human might want to drink.

The essence of vitality for any ecosystem is complexity and balance. In a free-flowing mountain river, the physical, chemical, and biological conditions that constitute habitat for a single living creature change drastically over short distances in all directions—upstream, downstream, shallower, deeper, in front of a rock, behind it, under it. This heterogeneity makes for spectacular diversity of species, comparable to an ocean shelf, or the heart of an equatorial rainforest. And that diversity in its turn makes for great complexity of interlocking relationships, great richness of life, and balance.

But the balance, in a river, is especially precarious, especially delicate: because the water never stops moving. The pressure never relents. The boundary between life and death is measured in millimeters. There is no room for error.

# A
# DEATHLY CHILL

*Hypothermia*
*in Fact and Supposition*

Death is personal. One of those things where you do it or you talk about it but not both: by its essence a conspiracy of silence. Notwithstanding Elizabeth Kübler-Ross, it is the single inevitable human enterprise that we can have no hope of comprehending experientially—not worth a damn, anyway—in advance. The contemplation and "comprehension" of death is, after all, something *live* people do, or try to, by linguistic bamboozlement of ourselves, chasing the end of a rainbow with assay tools. I repeat all this obvious stuff here because of the Ram Patrol of Chattaroy, Washington, and the question of hypothermia.

On the 10th of August, 1982, an Associated Press story ran in a corner of the back page of my newspaper under the headline: "Hypothermia Blamed in Deaths of Scouts." It was bizarre and pathetic. Four Boy Scouts and two adult leaders had been found drifting, dead, in a cove of a glacier-fed Canadian lake near the west border of Banff National Park. They were stragglers from a canoe trip that had included twenty-three other boys and men, and had been missing since just the previous afternoon. When discovered, all of the corpses were floating head-up and neatly strapped into life jackets, not far from their undamaged canoes.

There was no sign of accident, desperate struggle, or panic. In fact, one of the two adults was still wearing his glasses and hat. "They were in the Ram Patrol, our most experienced group," a scoutmaster told the AP. "We had them follow the others because they were the best." The water temperature was steady at around 45°F. and all of the scouts had been in and out of it, swimming and bathing, for the whole week. The presumed cause of death was hypothermia.

It would not be quite accurate to say that the four boys and two men of the Ram Patrol had frozen to death—not at 45°F. Rather, evidently, they had been *chilled* to death. "It was like they had just gone to sleep in the water," said the scoutmaster. "They probably ran out of energy and died." What seemed obvious to this man, by hindsight, had apparently stolen upon six robust young campers like a Mosaic angel of death. Given the nature of hypothermia, and the nature of water, it is not hard to believe.

Seventy years ago hypothermia, like radiation sickness, was unheard of. People in those days died of consumption, yellow fever, childbirth, the flu. They also, in cases of mishap on the high seas, died of "drowning." But drowning was merely the standard official and popular presumption, clung to for lack of a better one: live people doing their best, again, to get a grip on the lonely and personal business of death. As late as 1969 a physiologist from Oxford University, W. R. Keatinge, wrote that "until recently even experts commonly regarded drowning as the only important hazard to life in the water. Those who did look farther seldom appreciated any other hazards except thirst and attack by sharks. This belief is still common. It is almost routine for anyone who dies in the water to be said to have drowned, not only in everyday conversation and the press but often in official reports." For example, the case of the *Titanic*.

It was on its maiden trip when an iceberg hit the ship. This was just before midnight on April 14, 1912, a chilly spring night in the North Atlantic, and the temperature of the water in which the *Titanic* sank, so quickly, was hovering down around 32°F. During the early minutes of pandemonium roughly one-third of the total of passengers and crew managed to get safely aboard lifeboats, either dry-shod from the side of the ship itself or after a brief dunking. The other 1,489 people were left swimming, but there were far more than enough life jackets to take care of everybody. They weren't far from the English coast and within just one hour and fifty minutes another ship, the *Carpathia*, had arrived on the scene to begin scooping up survivors. Now the shocking part. The *Carpathia* was able to save almost all of those lucky or assertive folks who had gotten themselves a place in the lifeboats; *every one* of the other 1,489 people, most of them bobbing there wet in perfectly decent life jackets, was dead.

Afterward an official report came down from the Superintendent of the Port of Southhampton, under the title "Particulars relating to the deaths of members of the crew lately on board the S.S. *Titanic.*" This document included a roll of fatalities that ran nineteen pages long, and after each name the cause of death was cited as drowning. In all the various investigations and reports following the *Titanic* disaster, says Keatinge, there was hardly a mention of immersion hypothermia—under that phrase or any other—as a cause of or contributor toward death.

Nowadays scientists and maritime people know better. Perhaps no form of exposure to nature's brutal indifference is more deadly and (aside from running afoul a grizzly or toppling off El Capitan) more swift than hypothermia. Yet it is also insidiously subtle. About 600 Americans die from it each year, and despite the common notion that associates hypothermia with arctic cold, most of those 600 victims were never in any danger of frostbite. Many of them were subjected only to cool or even mild temperatures, in the 40s and 50s, but for one reason or another they got caught with wet clothing in the path of brisk winds, and couldn't protect themselves before it was too late. Many others were simply plunked into an ocean or a stream or a lake, like the Ram Patrol, and for one reason or another couldn't get out. They didn't freeze to death, like that smug dude in the Jack London story (which still represents the most widely known, and misleading, paradigm of fatal hypothermia). They chilled to death. It doesn't take long.

And the numbers are forbidding. Most clothing when it is wet (wool of course being the exception) loses up to 90 percent of its insulating value. Put a rain-drenched mountaineer on a breeze-raked ridge and, without a windbreaker, his life will be in jeopardy within half an hour. But full immersion in water, occupational hazard of sea travelers since the time of Noah, is more deadly still—because the thermal conductivity of water is 240 times that of still air. While a man overboard sculls gently to keep his face out of the waves, rides without further effort in his life jacket, waits hopefully for speedy rescue, the water sucks heat—and therefore life—out of the core of his body at an unbelievable rate. Immersed in water at 32°F., like the *Titanic* passengers, the average human will die within an hour. Immersed at 59°F., he will die after six hours. And 59° happens to be warmer than practically all of the coastal and inland waters of North America; in fact, it is warmer than most of the surface water on the planet ever gets (outside the tropics) through an entire year.

No wonder shipwreck, grand and small, has killed so many good swimmers. No wonder Madame Sosostris, famous clairvoyante, wisest woman in Europe, said: "Fear death by water."

As a body begins losing large amounts of heat to the surrounding environment, two things may happen. The less serious is frostbite, in which blood circulation to the extremities is automatically reduced as a desperate measure to conserve heat; this results in a drastic differential between the temperature of the skin and the temperature of the thoracic interior, and those expendable fingers and toes are sacrificed to maintain thermal stability in the body's vital core. The more serious is hypothermia, occurring when no such differential, no such desperate sacrifice, can prevent the core temperature itself from plummeting. As that core temperature falls, the symptoms of hypothermic trauma develop in progressive stages. A physician and mountaineer named Ted Lathrop, in a pamphlet published by the Mazamas climbing club, has described those stages in detail.

Dropping from a normal 98.6° down to 96° at the body core, says Lathrop, the victim will show uncontrollable shivering and a distinct onset of clumsiness. From 95° down to 91° the shivering will continue, and now speech will become slurred, mental acuity will decrease, there may also be amnesia. During this stage often come those crucial mistakes in judgment that prevent a victim from taking certain obvious steps that could save him from death. Between 90° and 86° the shivering will be replaced, says Lathrop, with extreme muscular rigidity, and exposed skin will sometimes appear blue or puffy. Mental coherence may be negligible, and amnesia may be total, though the person may still be able to walk and, if he is unlucky, his companions may not yet have noticed that he is in serious trouble. Down around 81° he will slide into a stupor, with reduced rate of pulse and respiration. Two or three degrees colder than that, at the body core, and he goes unconscious. His heartbeat may now be erratic, yet even if it remains steady there is a grave problem: Human blood cooled to this temperature becomes reluctant to turn loose the oxygen it's supposed to transport, so despite continued circulation, the brain and the heart muscle itself may be starved of the oxygen they require. If the core temperature falls further, below 78°, those brain centers that govern heartbeat and respiration will probably give out. There will be cardiac fibrillation—that is, the heart will be gripped with disorganized spasmodic twitching. Also now, pulmonary edema and hemorrhage—the lungs suddenly filling with clear cellular liquids and blood. The person may vomit, or just cough heartrackingly, a bit of pink foam frothing out from between his lips. And then he is dead.

The coldest spot of skin on his coldest toe may be no cooler than 59°F. But his core has fallen to 75°, and all the gears have seized.

The conditions in Lake McNaughton, where the Ram Patrol came to their end, were more than sufficiently inhospitable to bring on this sequence of stages, and to justify that surviving scoutmaster in his diagnosis. True enough, it would be hard to imagine how six healthy young men could drown under such circumstances, but not hard at all to figure how hypothermia might have killed them. The lake had gotten a bit choppy in late afternoon, and the Ram Patrol must have pulled into that cove for shelter, the scoutmaster guessed, when their canoes started taking on bilge. They seemed to have climbed out of the boats intentionally in shallow water, he guessed, and even succeeded in dumping both canoes empty, and righting them again. Or something. "Then they must have run out of energy and hypothermia set in," said the scoutmaster to the Associated Press.

But then later that week I made a call to the coroner of Revelstoke, British Columbia, within whose jurisdiction the Lake McNaughton incident occurred. The coroner told me something different. Sometime after the first newspaper story, autopsies were performed. Hypothermia, it had been quietly concluded, was *not* the cause of the deaths.

The lungs of the victims were filled not with blood and clear bodily fluids, but with lake water. The four boys and two men of the Ram Patrol, floating in life jackets near their empty canoes, had all drowned.

No one knows how. No one is likely to find out. Early symptoms of hypothermia, such as lassitude or muscular rigidity, may have made some secondary contribution, said the coroner. But that, he admitted, was purely speculative. With no witnesses and no survivors, the truth could only be guessed at. Death is personal.

# IS SEX
# NECESSARY?

---

*Virgin Birth
and Opportunism in the Garden*

*Birds do it, bees do it*, goes the tune. But the songsters, as usual, would mislead us with drastic oversimplifications. The full truth happens to be more eccentrically non-libidinous: Sometimes they *don't* do it, those very creatures, and get the same results anyway. Bees of all species, for instance, are notable to geneticists precisely for their ability to produce offspring while doing *without*. Likewise at least one variety of bird—the Beltsville Small White turkey, a domestic dinner-table model out of Beltsville, Maryland—has achieved scientific renown for a similar feat. What we are talking about here is celibate motherhood, procreation without copulation, a phenomenon that goes by the technical name *parthenogenesis*. Translated from the Greek roots: virgin birth.

And you don't have to be Catholic to believe in this one.

Miraculous as it may seem, parthenogenesis is actually rather common throughout nature, practiced regularly or intermittently by at least some species within almost every group of animals except (for reasons still unknown) dragonflies and mammals. Reproduction by virgin females has been discovered among reptiles, birds, fishes, amphibians, crustaceans, molluscs, ticks, the jelly-fish clan, flatworms, roundworms, segmented worms; and among insects (notwithstanding those unrelentingly sexy dragonflies) it is especially favored. The order Hymenoptera, including all bees and wasps, is uniformly parthenogenetic in the manner by which males are produced: Every male honeybee is born without any genetic contribution from a father. Among the beetles, there are thirty-five different forms of parthenogenetic weevil. The African weaver ant employs parthenogenesis, as do twenty-three species of fruit fly and at least one kind of roach. The gall midge *Miastor* is notorious for the exceptionally bizarre and grisly scenario that allows its fatherless young to see daylight: *Miastor* daughters cannibalize the mother from inside, with ruthless impatience, until her hollowed-out skin splits open like the door of an overcrowded nursery. But the foremost practitioners of virgin birth—their elaborate and versatile proficiency unmatched in the animal kingdom—are undoubtedly the aphids.

Now no sensible reader of even this book can be expected, I realize, to care faintly about aphid biology qua aphid biology. That's just asking too much. But there's a larger rationale for dragging you aphid-ward. The life cycle of these little nebbishy sap-sucking insects, the very same that infest rose bushes and house plants, not only exemplifies *how* parthenogenetic reproduction is done; it also very clearly shows *why*.

First the biographical facts. A typical aphid, which feeds entirely on plant juices tapped off from the vascular system of young leaves, spends winter dormant and protected, as an egg. The egg is attached near a bud site on the new growth of a poplar tree. In March, when the tree sap has begun to rise and the buds have begun to burgeon, an aphid hatchling appears, plugging its sharp snout (like a mosquito's) into the tree's tenderest plumbing. This solitary individual aphid will be, necessarily, a wingless female. If she is lucky, she will become sole founder of a vast aphid population. Having sucked enough poplar sap to reach maturity, she produces—by *live birth* now, and without benefit of a mate—daughters identical to herself. These wingless daughters also plug into the tree's flow of sap, and they also produce further wingless daughters, until sometime in late May, when that particular branch of that particular tree can support no more thirsty aphids. Suddenly there is a change: The next generation of daughters are born with wings. They fly off in search of a better situation.

One such aviatrix lands on an herbaceous plant—say a young climbing bean in some human's garden—and the pattern repeats. She plugs into the sap ducts on the underside of a new leaf, commences feasting destructively, and delivers by parthenogenesis a great brood of wingless daughters. The daughters beget more daughters, those daughters beget still more, and so on, until the poor bean plant is encrusted with a solid mob of these fat little elbowing greedy sisters. Then again, neatly triggered by the crowded conditions, a generation of daughters are born with wings. Away they fly, looking for prospects, and one of them lights on, say, a sugar beet. (The switch from bean to beet is fine, because our species of typical aphid is not inordinately choosy.) The sugar beet before long is covered, sucked upon mercilessly, victimized by a horde of mothers and nieces and granddaughters. Still not a single male aphid has appeared anywhere in the chain.

The lurching from one plant to another continues; the alternation between wingless and winged daughters continues. But in September, with fresh tender plant growth increasingly hard to find, there is another change.

Flying daughters are born who have a different destiny: They wing back to the poplar tree, where they give birth to a crop of wingless females that are unlike any so far. These latest girls know the meaning of sex! Meanwhile, at long last, the starving survivors back on that final bedraggled sugar beet have brought forth a generation of males. The males have wings. They take to the air in quest of poplar trees and first love. *Et voilà.* The mated females lay eggs that will wait out the winter near bud sites on that poplar tree, and the circle is thus completed. One single aphid hatchling—call her the *fundatrix*—in this way can give rise in the course of a year, from her own ovaries exclusively, to roughly a zillion aphids.

Well and good, you say. A zillion aphids. But what is the point of it?

The point, for aphids as for most other parthenogenetic animals, is (1) exceptionally fast reproduction that allows (2) maximal exploitation of temporary resource abundance and unstable environmental conditions, while (3) facilitating the successful colonization of unfamiliar habitats. In other words the aphid, like the gall midge and the weaver ant and the rest of their fellow parthenogens, is by its evolved character a galloping opportunist.

This is a term of science, not of abuse. Population ecologists make an illuminating distinction between what they label *equilibrium* and *opportunistic* species. According to William Birky and John Gilbert, from a paper in the journal *American Zoologist:* "Equilibrium species, exemplified by many vertebrates, maintain relatively constant population sizes, in part by being adapted to reproduce, at least slowly, in most of the environmental conditions which they meet. Opportunistic species, on the other hand, show extreme population fluctuations; they are adapted to reproduce only in a relatively narrow range of conditions, but make up for this by reproducing extremely rapidly in favorable circumstances. At least in some cases, opportunistic organisms can also be categorized as colonizing organisms." Birky and Gilbert also emphasize that "The potential for rapid reproduction is the essential evolutionary ticket for entry into the opportunistic life style."

And parthenogenesis, in turn, is the greatest time-saving gimmick in the history of animal reproduction. No hours or days are wasted while a female looks for a mate; no minutes lost to the act of mating itself. The female aphid attains sexual maturity and, bang, she becomes automatically pregnant. No waiting, no courtship, no fooling around. She delivers her brood of daughters, they grow to puberty and, zap, another generation immediately. If humans worked as fast, Jane Fonda today would be a great-grandmother. The time saved to parthenogenetic species may seem trivial, but it is not. It adds up dizzyingly: In the same time taken by a sexually reproducing insect to complete three generations for a total of 1,200 offspring, an aphid (assuming the *same* time required for each female to mature, and the *same* number of progeny in each litter), squandering no time on courtship or sex, will progress through six generations for an extended family of 318,000,000.

Even this isn't speedy enough for some restless opportunists. That matricidal gall midge *Miastor*, whose larvae feed on fleeting eruptions of fungus under the bark of trees, has developed a startling way to cut further time from the cycle of procreation. Far from waiting for a mate, *Miastor* does not even wait for maturity. When food is abundant, it is the *larva*, not the adult female fly, who is eaten alive from inside by her own daughters. And as those voracious daughters burst free of the husk that was their mother, each of them already contains further larval daughters taking shape ominously within its own ovaries. While the food lasts, while opportunity endures, no *Miastor* female can live to adulthood without dying of motherhood.

The implicit principle behind all this non-sexual reproduction, all this hurry, is simple: Don't argue with success. Don't tamper with a genetic blueprint that works. Unmated female aphids, and gall midges, pass on their own gene patterns virtually unaltered (except for the occasional mutation) to their daughters. Sexual reproduction, on the other hand, constitutes, by its essence, genetic tampering. The whole purpose of joining sperm with egg is to shuffle the genes of both parents and come up with a new combination that might perhaps be more advantageous. Give the kid something neither Mom nor Pop ever had. Parthenogenetic species, during their hurried phases at least, dispense with this genetic shuffle. They stick stubbornly to the gene pattern that seems to be working. They produce (with certain complicated exceptions) natural clones of themselves.

But what they gain thereby in reproductive rate, in great explosions of population, they give up in flexibility. They minimize their genetic options. They lessen their chances of adapting to unforeseen changes of circumstance.

Which is why more than one biologist has drawn the same conclusion as M. J. D. White: "Parthenogenetic forms seem to be frequently successful in the particular ecological niche which they occupy, but sooner or later the inherent disadvantages of their genetic system must be expected to lead to a lack of adaptability, followed by eventual extinction, or perhaps in some cases by a return to sexuality."

So it *is* necessary, at least intermittently (once a year, for the aphids, whether they need it or not), this thing called sex. As of course you and I knew it must be. Otherwise surely, by now, we mammals and dragonflies would have come up with something more dignified.

# DESERT
# SANITAIRE

The Englishman T. E. Lawrence, him of Arabian fame, was supposedly once asked why he so loved the desert. Skeptical historians have treated the great white-robed Lawrence less kindly than Peter O'Toole did, suggesting that the man was, perhaps not the reluctant demigod and charismatic camel-riding chieftain of Arab revolt as so appealingly pictured, but more on the line of a conniving, ambitious, self-mythologizing, pederastic, and congenitally mendacious poseur. Also, he stood only five foot three. To hear it from some biographers, the truth was not in him. He was his own press agent. He duped Lowell Thomas into spreading the legend which he himself had concocted, a legend more than faintly racist in its Kiplingesque presumption. He was a sadomasochistic neurotic whose entire life, say the critics, was "an enacted lie." But all this is just too harsh. Lawrence certainly had—he must have had—something every bit as precious as full mental health, or the habit of veracity: He had panache. He had high style. Had the gift for capturing, if not strict autobiographical truth, at least the human imagination. *Why do you love the desert?* they asked him, later, when he was languishing through his self-imposed obscurity back in soggy mucky England. Supposedly Lawrence-formerly-of-Arabia said: "Because it's clean."

At least I hope he did. Of course by a literal reading the notion is nonsense. Clean of what, dirt? Not if dust and perspiration and a week's funky unwashed body grime can be counted. Clean of microbial infestation and sinister many-legged vermin? Hardly. Perhaps clean of *human* infestation? That's more plausible, considering Lawrence's disposition. Anyhow, if you have ever spent time out there, not in Arabia necessarily but in the desert, down on the very ground, crunching off the miles with your boots, maybe you understand something of what poor troubled T.E. was getting at.

*It's clean.* It's austere. It's ascetic. Harshly unfertile and fatally inhospitable. Solitary. Unconnected. It's notable chiefly for what it lacks. America's own preeminent desert anchorite seems to agree: Wherever his head and feet may go, says Ed Abbey, his heart and guts linger loyally "here on the clean, true, comfortable rock, under the black sun of God's forsaken country." *It's clean.*

But what *is* it, this thing of such noteworthy cleanliness? "There is no single criterion," according to the renowned desert botanist Forrest Shreve, "by which a desert may be recognized and defined." Still we have to start somewhere. And a desert is one of those entities, like virginity and sans serif typeface, of which the definition must begin with negatives.

In this case, lack of water. Not enough rain. Less than ten inches of precipitation through the average year. A desert is not, most essentially, a hot place or a flat sandy place or a place filled with reptiles and cactus and dark-skinned people wearing strange headgear. Fact number one is that it's a dry place. Joseph Wood Krutch has written that "in desert country everything from the color of a mouse or the shape of a leaf up to the largest features of the mountains themselves is more likely than not to have the same explanation: dryness." From such a simple starting point, things get more complicated immediately.

The matter of sheer dryness, for instance, is less crucial than the matter of *aridity*, which is a measure of how much or how little water remains available on a particular landscape surface for how long. Ten inches of rain distributed evenly throughout a lengthy cool season will support plants and animals in modest profusion; ten inches dumped from a great cloudburst on one summer afternoon, then not another drop for the rest of the year, will produce a few hours of wild flooding and leave behind a typical parched desert, with wide empty arroyos and a scattering of peculiarly specialized creatures. Whatever water there may be comes and goes quickly in a desert, erratically, never remaining available over time. It abides not. It pours off the slopes of treeless mountains. It gathers volume in drywashes and roars peremptorily away. It soaks down fast through the sandy soil. Gone. Most of all it evaporates.

That's the other prerequisite for any desert environment, lesser partner to dryness: evaporation, as wrought by heat and wind. A little rain falls occasionally, yes, but coming as it does in prodigal storms during the warmest months, burnt off by direct sunshine and sucked away by the winds, the stuff disappears again almost at once. A system of land classification devised by Vladimir Köppen takes this into account, with a mathematical formula by which temperature and precipitation are together converted to an index of aridity. According to the Köppen method, any region where *potential evaporation* exceeds actual precipitation, by a certain margin, can be considered a desert. This rules out frigid locales with scant annual precipitation but plenty of permanent ice, such as Antarctica. Most of our own Southwest qualifies resoundingly.

But what, in the first place, makes a spot like Death Valley or Moab or Organ Pipe Monument so all-fired dry? Or for that matter a huge region like the Sahara? Or the Kalahari? Or the Taklimakan Desert of western China? Is it purely fortuitous that one geographical area—say the Amazon basin—should receive buckets of moisture while another area not far away—the

Atacama-Peruvian Desert—gets so precious little? The answer to that is no. Not at all fortuitous. The three different geophysical factors combine, generally at least two in each case, to produce the world's various zones of drastic and permanent drought: (1) high pressure systems of air in the horse latitudes, (2) shadowing mountains, and (3) cool ocean currents. Together these three even cast a tidy pattern.

Our planet wears its deserts like a fat woman in a hot red bikini. Don't take my word for it: Look at a globe. Spin the Earth and follow the Tropic of Cancer with your finger as it passes through, or very near, every great desert of the northern hemisphere: the Sahara, the Arabian, the Turkestan, the Dasht-i-Lut of Iran, the Thar of India, the Taklimakan, the Gobi, and back around to the coast of Baja. Now spin again and trace the Tropic of Capricorn, circling down there below the equator: through the Namib and the Kalahari in southwestern Africa, straight across to the big desert that constitutes central Australia, on around again to the Atacama-Peruvian and the Monte-Patagonian of South America. This arrangement is no coincidence. It's a result, first, of that high-pressure air in the horse latitudes.

The horse latitudes (traditionally so-called for tenuous and uninteresting reasons) encircle the Earth in two wide bands, one north of the equator and one south, respectively along the tropics of Cancer and Capricorn. The northern band spans roughly the area between latitudes 20°N and 35°N, and the counterpart covers a similar area of south latitudes. Between those two bands is the tropics, very hot and very wet, where most rainforest is located. That equatorial region is also the part of the terrestrial surface that—because of its distance from the poles of rotation—is moving with greatest velocity as the Earth spins through space. (The equator rolls around at better than 1,000 mph, while a point near the North Pole travels much slower.) For climatological reasons only slightly more obscure than Thomistic metaphysics, this differential in surface velocity produces trade winds, variations in barometric pressure, and a consistent trend of rising air over the tropics. As the air rises it grows cooler, therefore releasing its moisture (as cooling air always does) in generous deluge upon those tropical rainforests. Now the same air systems are high and dry: high aloft in the atmosphere, and emptied of their water. In that condition they slide out to the horse latitudes, north and south some hundreds of miles, and then again descend. Coming down, they compact themselves into high pressure systems of surpassing dryness. And as the pressure of this falling air increases, so does its temperature. The consequence is extreme permanent aridity along those two latitudinal bands, and a first cause for all the world's major deserts.

The second cause is mountains—long ranges of mountains, sprawling out across the path of prevailing winds. These ranges block the movement of moist air, forcing it to ascend over them like a water-skier over a jump. In the process that air is cooled to the point where it releases its water. The mountains get deep snow on their peaks and the land to leeward gets what is left: almost nothing. Such a *rain shadow* of dryness may stretch for hundreds of miles downwind, depending on the height of the range. No accident, then, that the Sahara is bordered along its northwestern rim by the Atlas Mountains, that the Taklimakan stares up at the Himalayas, that the Patagonian Desert is overshadowed by the Andes.

Ocean currents out of the polar regions work much the same way, sweeping up along the windward coastlines of certain continents and putting a chill into the oncoming weather systems before those systems quite reach the land. Abruptly cooled, the air masses drop their water off the coast and arrive inland with little to offer. For instance, the Benguela Current, curling up from Antarctica to lap the southwestern edge of Africa, steals moisture that might otherwise reach the Namib. The Humboldt Current, running cold up the west coast of South America, keeps the Atacama similarly deprived. And the California Current, flowing down from Alaska along the Pacific coast as far south as Baja, does its share to promote all-season baseball in Arizona.

Beyond all these causes of dryness, another important factor is wind, helping to shape desert not only through evaporation but also—and more drastically than in any other type of climatological zone—by way of erosion. Powerful winds blow almost constantly into and across any desert, with heavier cold air charging forward to fill the vacuum as hot light air rises away off the desert floor. Desert mountains tend to increase this gustiness, and in some cases to focus winds through canyons and passes for still more extreme effect. In desert they call the wind *chubasco* if it's a fierce rotary hurricane of a thing, whirling up wet and mean out of the tropics and tearing into the hot southern drylands with velocities up to 100 mph, sometimes delivering more than a year's average rainfall in just an afternoon. More innocent little whirlwinds, localized twisters and dust devils, are known as *tornillos*. The steadiest and driest wind goes by *sirocco*. Besides raking away moisture and making life tough for plants and animals, the wind works at dismantling mountains, grinding rock fragments into sand, piling the sand into dunes and moving them off like a herd of sheep. Uwe George has called desert wind "the greatest sand-blasting machine on earth," and there is vivid evidence for that notion in any number of desert formations.

The winds and the flash floods are further abetted, in trashing the terrain, by huge fluctuations in surface temperature. A desert thermometer doesn't just go way up, but wildly up and down, by day and by night, because those clear skies and that lack of vegetation let nearly all the day's solar energy radiate back away after dark. Easy come, easy go: In the desert air there is no insulation to slow the transfer of heat. And the temperature of the land surface fluctuates even more radically than the air temperature—a dark stone heated to 175°F. in the afternoon may cool to 50°F. overnight. The result is a constant process of fragmentation: rocks splitting themselves into pieces, with a sound like a gunshot, as though from sheer exasperation.

It is all so elaborately and neatly interconnected: The dryness of desert regions entails clear skies and a paucity of plants; which together entail fierce surface heat by day, bitter chill by night; which leads to rock fracture, crumbling mountains, and the eventual creation of sand. The thermal convection of air brings strong winds, which exacerbate in their turn the aridity and the erosion; also the irregularity of rainfall, acting upon soil not anchored by a continuous carpet of plants, creates the arroyos, the canyons, the badlands, the rugged bare mountains; wind and sand collaborate on the dunes and the sculpted rocks. Add to this a team of small thirst-proof mammals like the kangaroo rat, hardy birds like the poorwill, ingeniously appointed reptiles like the sidewinder, arthropods of all unspeakable and menacing variety—and what you have is a desert. A land of hardship, of durable living creatures but not many, of severe beauty, and in some ineffable way, of *cleanliness*.

In *Seven Pillars of Wisdom*, Lawrence wrote of the desert-dwelling race of human who had "embraced with all his soul this nakedness too harsh for volunteers, for the reason, felt but inarticulate, that there he found himself indubitably free. He lost material ties, comforts, all superfluities and other complications to achieve a personal liberty which haunted starvation and death. He saw no virtue in poverty herself: he enjoyed the little vices and luxuries— coffee, fresh water, women—which he could still preserve. In his life he had air and winds, sun and light, open spaces and a great emptiness. There was no human effort, no fecundity in Nature: just the heaven above and the unspotted earth beneath. There unconsciously he came near God."

Lawrence was talking about the Bedouin, but it might apply just as well to mad dogs and Englishmen. No cool distant tone of the anthropologist in those sentences, but an intimacy that sounds autobiographical (except for the condescending fillip to women) and more than a little nostalgic. Maybe that's what Thomas Edward Lawrence meant with his notion of the desert's *cleanliness:* For him, in some way, it had been next-to-Godliness.

# THE MIRACLE
# OF BLUBBER

---

*How the Pinnipeds
Avoid Meltdown in the Arctic*

A Norwegian who had transplanted himself to upper Baffin Island once told me his secret for sleeping warm on the winter tundra eighty miles north of the Arctic Circle. We were both in the audience of a lecture on winter-camping technique and this scrubby Norski had grown impatient with all the jaw-flapping about exotic lightweight tents, about goose-down booties, about sleeping bags with triple sandwich construction and differential layering and self-repairing nylon zippers and offset cross-baffle seams, not to mention the whole realm of PolarGuard and GoreTex and Ensolite and Thinsulate. It was all so much Samsonite luggage to him. He did not bother with a tent. Nor an igloo nor a snow cave. His bag was relatively cheap and unsophisticated, and he just laid it out on the open ground sleeved in a long plastic tube, like an extended Hefty trash liner, to cut the wind. But that wasn't the heart of his method. This was: Inside the bag he stripped naked, then smeared his body with a mixture of fish oil and Bag Balm, the noted unguent for chafed cattle udders, which he carried pre-mixed in a small jar. Thereby insuring total toasty comfort—he claimed—at 40 below with a breeze. It was inexpensive, fail-proof, and quite portable. But were there no disadvantages? I asked. Actually yes, there was one.

"The smell," he told me.

We can all probably think of a few others, but that's not the point. The point, for purposes here, is that this fellow's solution to the problem—how to stay warm in the coldest places on Earth—was very much in the same spirit as the one used by an order of incongruous mammals called the *Pinnipedia*.

The pinnipeds are those most whole-heartedly Arctic of all warm-blooded animals, the seals and sea lions and walruses. Linnaeus said of them: "This is a dirty, curious, quarrelsome tribe, easily tamed, and polygamous; flesh succulent and tender; fat and skin useful. They inhabit and swim under water and crawl on land with difficulty because of their retracted fore-feet and united hind-feet; feed on fish and marine productions, and swallow stones to prevent hunger, by distending the stomach." His portrait is a bit crankish, and underestimates the terrestrial mobility of sea lions and walruses, but generally Linnaeus was accurate—they are desperate characters, these pinnipeds, low on decorum and stubbornly heterodox. While the musk ox and the snowy owl and other northerly creatures were evolving fine-tuned physiological tools and elaborate behavioral adaptations that allow them to cope marginally with extreme cold, the pinnipeds hit on an answer that is as simple, as perfunctory, and as inelegant as the greased Norwegian's: blubber.

For insulation against heat loss, the pinnipeds rely on blubber. As an energy supplement during times when food is scarce, they metabolize their stored blubber. Giving them high buoyancy in the water, blubber; padding out their roundish shapes to reduce surface area and in that way further minimize the escape of their body heat, blubber; for hydrodynamic streamlining, blubber. The stuff has more uses than Dr. Bronner's soap.

A few species of pinniped prefer coolish temperate waters, some are native to the fringes of the Antarctic continent, several untypical forms even sneak into the tropics, but most of the pinnipeds spend their entire lives frolicking near and on and under the great platform of ice that covers the Arctic Ocean, diving for food in super-cooled seawater that is never much warmer than 30°F., basking dry in air that is often 70 degrees colder, mating and bearing their young on the ice, migrating not, hibernating not, doing business throughout the bitterest cold of the long winter night. They suck in a breath and stay underwater for fifteen minutes at a stretch, their metabolism shifting down as they chase fish or root up clams, so that they burn *less* oxygen during this exertion than when they are topside. They chew breathing holes through the new ice in autumn, a network of holes proprietary to each animal, each hole just large enough to let them poke a snout up for air, and they keep the holes open with more chewing as winter thickens the ice to three or four feet. They perform other marvels of hardiness, all the while wrapped in a thick layer of blubber, a natural wetsuit, from which only the head and the flippers protrude, like sprouts on a restless potato. The name *pinniped* means "wing-footed," or alternately "fin-footed," according to which translation you choose, but either will do, since these blithe creatures move through icy water like fish, or alternately, like the messenger-god Hermes. Notwithstanding blubber.

Blubber consists mainly of lipid oils but to say "blubber is fat," and leave it there, would be roughly as accurate as explaining that pâté de foie gras is chopped liver. Blubber arises in the hypodermis, a layer of skin between the dermis and the thin sheet of muscle that controls skin movement. The blubber itself is a matrix of snaky protein strands (collagen fibers) randomly intertangled, with the spaces among them filled by fat cells. The fat cells account for a much larger portion than the collagen fibers, which provide just the minimal necessary support. A few nerves and a sizable number of ingeniously arranged blood vessels also penetrate this layer. The particular arrangement of small arteries and veins is instrumental in giving the pinnipeds their prodigious and versatile powers of regulating heat loss. More on this presently.

Herman Melville tells us that "blubber is something of the consistence of firm, close-grained beef, but tougher, more elastic and compact, and ranges from eight or ten to twelve and fifteen inches in thickness," but of course he is talking about a sperm whale, and most of the whales are far over-equipped for any chilly waters they ever visit. Their foot-thick blubber serves more use as an energy reservoir during long migrations than as insulation. An adult walrus has blubber ranging two to four inches thick, and even a ringed seal, one of the smaller Arctic species, might have a uniform wrap of nearly two inches. If you freeze a dead seal solid and then cut him in half with a band saw (one relentless physiologist has done this, thereby setting a record of some sort for barbarous scientific literal-mindedness), in cross-section the poor seal with his blubber looks something like a Hostess Twinkie.

That layer of blubber is what makes possible a largely aquatic life in an Arctic environment. Most Arctic mammals are insulated with dense fur, not blubber, because most Arctic mammals (excepting the pinnipeds and the polar bear) have evolved under the premise that they will do their utmost to avoid getting wet. Water has a drastically greater chilling capacity than still air at the same temperature; seawater as cold as 30° F. can draw heat away faster than a harsh wind on a sub-zero mountain peak; and fur generally offers far less insulation value in a watery medium than in air. A fur-bundled Eskimo so unfortunate as to fall through the ice will freeze to death in less than half an hour. But when a seal or walrus plunges into the same water, some fancy adjustments occur quickly in the animal's blubber, and heat loss is kept to a minimum.

Those adjustments are known technically as "the heterothermic operation of a homeotherm." This simply means that the blubbery pinnipeds, though they are warm-blooded and must maintain their internal body temperature at around 100°F., are nearly immune to frostbite. They protect their heart and brain and other vital organs from disastrous chilling precisely by letting their skin, and the extremities of their flippers, grow very cold.

The skin of a ringed seal in icy water, for instance, may routinely cool down to 33°F., only a few degrees warmer than the sea itself and 67 degrees colder than the animal's core temperature. As long as the outermost blubber and the epidermis that covers it are barely warmer than the temperature at which the fluids of those cells will freeze solid, they remain healthy; blubber can manage that trick, whereas most animal tissue would be wrecked; meanwhile, heat loss from cold skin to cold ocean will be insignificant. This internalized thermal gradient, crucial to pinnipeds, is all accomplished by that ingenious arrangement of blood vessels.

The trick occurs near the surface of a seal's blubber, where the arteries carrying warm blood from the heart begin branching down into smaller arterioles—which will branch down further to tiny capillaries, in which the blood finally makes its U-turn into veins that lead back toward the heart. Some of those arterioles are set in very close proximity to the small veins. This cozy juxtaposition of outgoing and incoming vessels is called the "vascular countercurrent heat-exchange system," and accounts for the pinnipeds' ability to keep their outermost blubber supplied with blood but starved of heat. The blood goes out and around through the capillaries, but the heat takes a shortcut: It diffuses from the arterioles to the nearby veins. So it never gets near enough to the surface to escape from the animal's body.

And on a warm summer day—the sort of Arctic scorcher, say 40°F. in the sun, that threatens furry sea otters with heat prostration and can curdle the blood of a beached whale—the exchanger vessels in a seal's blubber are simply bypassed, by autonomous constriction and dilation, and heat is mercifully released from the animal's capillaries. You can't do that with fur or with down. You can't even do that with PolarGuard.

This countercurrent heat-exchange system happens to be the same setup by which, in a nuclear power reactor, heat is transferred from the fission core to the steam turbines, while radioactivity is (theoretically) prevented from escaping the containment building. But don't hold that against the pinnipeds. The main difference between the heat-control mechanism in a ringed seal, and the one used at Three Mile Island Unit Two, is of course that the seal's works.

It works beautifully. Pinnipeds may be killed by sharks and polar bears. With bad luck they may suffocate under the ice or be speared by an Eskimo as they bob up for air. Occasionally one may be smashed by storm-driven waves, or shifting ice floes, or a clumsy bull walrus. But very seldom, we can presume, does an adult Arctic seal freeze to death.

The wonders of blubber have been documented by science. And me, I still haven't tested fish oil and Bag Balm.

# DEAD THINGS
# IN THE WATER

---

*Wherein Whisperin' Jack*
*Goes Hungry But Helps Solve a Riddle*

A vigorous mountain river will have just its proper share of green-ish slimes, filamentous botanical cruds, tiny crawling and squiggling critters of unimaginable grotesque appearance, sleek trout flashing discreetly and, prerequisite to all, dead things in the water.

Some mountain rivers, on the other hand, contain almost nothing but water: cold and pure and limpid. A pitiful state of affairs. They can be pretty to look at, these sterile streams, and refreshing to drink from, and excellent for floating a kayak, yet they lack the network of exotic life forms that constitute a rich riverine ecosystem. This condition is most common in the extreme headwaters of glacier-fed rivers, where the cascading meltwater has splashed its way downstream only a mile or two from the glacial source. But the same sort of pristine sterility can also exist, under certain circumstances, far from the source and downstream from fertile stretches. There happens to be one such zone, maybe three miles long, on the South Fork of the Flathead River in northwestern Montana. I couldn't have known that, of course, not from the maps or the guidebooks, when I arranged to go there with Whisperin' Jack. Otherwise I might have taken some food.

Whisperin' Jack is six foot five and weighs about 140 and wears a brown Dobbs fedora that, despite his degree from the Harvard Medical School, makes him look like the kind of quiet creepy guy whose car trunk is one day discovered to contain the sucked-upon fingerbones of missing hitchhikers. He can be a wonderful expeditionary companion, Whisperin' Jack, but he remains nevertheless a city lad, so when you get him lost in the dark on cross-country skis or collapse a snow cave on his head or capsize him in foamy water, he is liable to act just a little whiny. He watched me from hollowed eyesockets while I tossed four small cans of barbecued herring and a package of dried soup into the pack, but I promised him we would have plenty to eat. *See, Jack, I'm taking a pound of butter. See, two lemons.* The assumption, a hubristic one as it turned out, was that I would catch trout. Also in my pack was a fly rod and some plastic boxes holding 8,000 scientifically rendered imitation aquatic insects—though I might just as well have been carrying lobster bibs and a pair of nutcrackers, for all the good that fishing gear would do us.

Our destination was a stretch of the upper South Fork of the Flathead, barely outside the north boundary of the Bob Marshall Wilderness, and our chief purpose, mine at least, was to catch and examine and release a few samples of America's newest species of salmonid, *Salvelinus confluentus*, commonly called the bull trout. This creature has been around for a long time, but it was always considered a strain of Dolly Varden trout (which is itself technically not a trout but a char) until several years ago, when ichthyologists concluded on new evidence that it should be recognized as a distinct species, indigenous mainly to Montana. The point of interest for me was not taxonomic hair-splitting, but sport. Besides being mildly newsy, *S. confluentus* is aggressive, voracious and, by trout standards, huge: A rainbow trout weighing five pounds is heroic, but bull trout of ten and twelve pounds are not unusual. A September trek up the South Fork—where *confluentus* from far downstream go, in autumn, to spawn—therefore seemed justified. The river was also reportedly teeming with cutthroat trout in sizes that would fit a skillet.

And just a couple miles upstream from where Jack and I went, it is: teeming with cutthroat and the insects they eat and the microorganisms on which those insects feed. A couple miles downstream? Teeming there too. Lots of teeming everywhere but the Bunker Creek gorge, to which in our ignorance Whisperin' Jack and I duly marched. We left the trail and climbed down a steep rocky face, 200 feet, to where the river sat beautifully pinched at the bottom of this gloriously protective gorge. And found nothing but water: cold and pure and limpid. We had come to a zone of pristine sterility.

For three days I fished hard, in fine deep pools and classic riffles, without ever a strike; for three days we ate barbequed herring and dried soup. Then Whisperin' Jack ate the butter. I tried all 8,000 flies but that was wasted motion, because after an hour on the first evening I was sure we would go skunked. I knew it from staring into the water. I just hadn't yet figured out why.

The cycle of energy flow among the life forms of a river ecosystem must begin, oddly, with dead things in the water. The experts in their technical jargon call this stuff *allochthonous organic material*, by which they mean, mainly, the routine sheddings of nearby terrestrial plants. A rainforest river like the Amazon may fuel its energy cycle with falling blossoms, fruit, and pollen from trees along the bank; a mountain river like the Flathead does it, every autumn, with a vast dumping of leaves. These dead leaves, blown onto the water from cottonwoods and aspen and willow and alder, contain large concentrations of protein that is usable by aquatic fungi and bacteria. The leaves also release dissolved nitrate, a nutrient that enhances production of algae across the stream bottom. And together the fungi, the bacteria, and the various forms of algae constitute the first level of productivity in the river's food chain, the category of tiniest and simplest living organisms on which all the others depend.

During late autumn and winter the fallen leaves become sodden, settling to the bottom, collecting in backwaters, wedging in

clots between rocks. Bacteria and fungi climb aboard, begin feasting, digesting, causing decay, until eventually it all crumbles to
a fine compost that sails away on the current. Such rotting plant
debris is called *detritus*. It is the manna of rivers.

Downstream a menagerie of bizarrely shaped insects, unlike
anything you've ever seen on dry land, scoop in this manna with
all fashion of ingenious equipment and strange anatomical ap-
paratus—silken webs like a wind sock, lower lips like a cranberry
rake—and devour it, bacteria and fungi included. These detritus-
eaters are the larval stages of certain caddisflies and stoneflies and
midges and mayflies, breathing through gills and making a hard
living among the crevices of the riverbed. Along with related
species of the same insect groups—species that prefer grazing
algae off rocks—they are the primary consumers in a river ecosystem, the creatures responsible for turning vegetable matter
into meat.

Preying upon those herbivores are the secondary consumers:
carnivorous insects slightly bigger and more fierce, predatory
species of stonefly and caddisfly and dragonfly and beetle. Put a
seine in the water of a fertile mountain river, turn over a few
rocks, and you can lift out handfuls of these amazing beasts, some
up to two inches long, clambering harmlessly over your cupped
palms like surreal variations of miniature six-legged dinosaur. On
their best days they rule as lords of the streambed jungle; but
one bad day for a large stonefly, one careless moment of exposure,
and it is in the belly of the tertiary consumer, the trout.

Each of these levels of energy transformation—from the cottonwood leaf, through the fungi and the mayflies and all the rest,
eventually to the trout—is necessary to every other level suc-
ceeding it. Algae can flourish in the absence of caddisflies, caddisflies can live without trout, but not vice versa. With one link
removed, the rest of the chain disappears.

And eureka. On the third day of our diet camp on the South
Fork I was pondering just these facts, while staring into the pellucid water, waving a fly rod futilely through the air, towered

over by the walls of the Bunker Creek gorge, when suddenly I understood. Back to the campsite I stumbled, now in high spirits. Whisperin' Jack was sitting against a bleached log.

"Good news, Jack. I've figured it out. I know why we aren't catching fish."

"Me too. Incompetent guide."

"Wrong," I said. "It's because there are no dead things in the water."

"We could fix that."

"No detritus. Not a scrap of it, not a wisp, not a shred. The backwaters and crannies should be full of leaf material a year or two old. But no, nothing. I haven't seen any dead vegetation that's been in the water more than a couple months. Spring runoff must come through this gorge like Montezuma's revenge. Purging out all the plant debris, see, before it can decay. Before it can be fed on through a summer. So the river, from bottom to top, is just too clean. No leafy funk, no little germy guys, no insects, no trout." All this, I'm afraid, in a tone of epiphanic jubilation. "The riddle is solved."

"That's great, Sacajawea," said Whisperin' Jack. He still wasn't smiling. Then he ate the lemons.

# YIN AND YANG
# IN THE TULAROSA
# BASIN

In the outback of southern New Mexico, laid down across a thousand square miles of otherwise unexceptional desert, there is a message.

It is gigantic and stark. A simple design, vaguely familiar, executed in unearthly black-and-white against the brown desert ground. You could see it from the moon with a cheap telescope. Nevertheless it is more cryptic than Stonehenge. An army of zealous Chinese stonemasons might have spent their lifetimes erecting such a thing—but no, that's not where it came from. The full design includes four elements, only two of those man-made, the others attributable to natural processes of the sort that are loosely called "acts of God." In the desert, as Moses found out, God tends to act with an especially bold hand. The constituent materials in this particular case are basalt rock and gypsum sand; black and white, rough and smooth, hard and soft; dissimilar as fire and ice. This message is drawn in high contrast. The text is clear but the meaning is not.

A dune field of startling whiteness, called the White Sands, sprawls out in giant ameboid shape, creeping northeastward with the winds. A hardened black flow of recent volcanic lava, called the Carrizozo Malpais, stretches southwestward down the gentle incline along which gravity poured it. Near an old ranch site known as Three Rivers, the lead edge of the whiteness approaches the forward lip of blackness, across a gap of not many miles. Through that gap runs a narrow range road, open for public travel only one day a year. Where the road goes, why armed men in guardhouses monitor its disuse, are questions for later. The oddities about this place will have to be taken in turn. We are in the Tularosa Basin, a sunken valley full of saltbush and lizards and history, gypsum and lava, more than its share of preternatural resonance, lying halfway between Las Cruces and Roswell on the way to nowhere at all. We are here, first of all, for the big design.

The design: Laid out across the desert by a convergence of geological accidents, it is an unmistakable yin-and-yang, a huge version of that Taoist emblem that stands for the paradox of dialectical oneness—two teardrops bound complementarily into a circle, dark and light, head to tail, representing the unity within which all worldly flux remains balanced. In this particular case, the emblem itself is as large as Long Island. From the moon or beyond, with your telescope, you might take it for a signal of harmony. The confusion would be understandable.

An abundance of gypsum was the earliest of those geologic accidents.

Gypsum is curious stuff from which to make a dune field. In scientific notation it is $CaSO_4 \cdot 2H_2O$, meaning simply the mineral calcium sulphate, bound up in crystalline form with a proportion of plain water. More familiarly, it is the main ingredient of plaster of Paris. Under ideal conditions, falling out of a heavy solution, it grows into elegant dagger-like crystals called selenite, which are more or less clear or amber depending on purity. But as erosional forces break selenite down into small granules—sand—the faces of those granules, being relatively soft, become scratched.

The scratches scatter light. The result is whiteness. You hear the name White Sands but, until you make your own pilgrimage, until you lose yourself in the heart of these dunes with only a canteen and a compass, the words are unlikely to register as they should.

Take them literally. White sands. Whiteness like ivory. Like sun-bleached skull of lost desert cow. Whiteness like January in the Absaroka Mountains between Montana and Wyoming, 200 yards above timberline. Actually there is nothing and nowhere else quite like White Sands, the world's largest expanse of wind-blown gypsum. Nowhere else on Earth where you can surround yourself with such profound whiteness, and still be in danger of snakebite. The white dunes began forming perhaps 25,000 years ago, but the gypsum itself has been here much longer.

It was deposited during the late Paleozoic era, gypsum-rich layers of sedimentary rock left behind from gypsum-rich sea waters as the long cycles of climate moved an ocean coast back and forth over what is now southern New Mexico. Other sediments were left in the course of other cycles, burying the gypsum beneath hundreds of feet of limestone and shale. It might have stayed there—inert and hidden, like most of the gypsum in the Earth's crust—if not for the next accident. Geologic pressures that were creating the Rocky Mountains also caused this particular area to buckle gently upward into a high rounded plateau. Roughly 10 million years ago, another shifting of pressures caused a pair of fault lines to develop, running north-south for a hundred miles; and along these faults, the plateau fell like a cake. The parallel fault lines became a matched set of continuous mountain cliffs, facing each other over a sunken basin. On the east side looking west, the Sacramento range; on the west looking east, the San Andres. In between was the Tularosa.

To the north and south, also, the Tularosa was blocked by high ground. Like many valleys in the desert country of the West, it had no outlet to the sea. So when the next cycle of wetness began, this basin turned into a vast lake.

Erosional torrents flowed down from the mountains to fill it, and (because the mineral is easily eroded and highly soluble) those waters carried gypsum. The lake in its turn became gypsum-rich. Then our most recent age of relative drought dried it away to almost nothing. The waters shrank gradually back to the lowest spot in the basin, a small area toward the southwest corner, and as the big lake gave up the ghost, it also gave up the gypsum. Meanwhile groundwater flow from the upper end of the basin also carried dissolved gypsum, underground, toward the same spot. These days Lake Lucero is never more than a briney puddle, shin-deep at the end of the monsoon season. For most of the year, it is only a dry bed. But it derives a mute dignity from being the source of the White Sands.

With each cycle of evaporation—now, as for thousands of years—small selenite crystals bloom magically along the puddle's margins. Seasonal windstorms roar out of the southwest, through gaps in the San Andres mountains, grinding the crystals to sand. And so the dunes gather themselves, rise, and move.

The nimblest of them advance about thirty feet in a year. Others travel more slowly. Today the sand is spread over an area of almost 300 square miles, enough to constitute a distinct ecosystem with its own patterns of organismic association, its (temporarily) stabilized zones of vegetation, its uniquely adapted races of animal. A whole world of life hides at the heart of the White Sands. But despite 25,000 years, there is a gypsy quality to that life. The restlessness of the dunes imposes special demands. The very ground here is in motion. *We are the dunes: we cover all. You must move along with us, or get out of the way, or die.* The animals, even the plants, manage to cope with that imperative in their own patient ways.

The White Sands are gliding northeastward, inexorably, toward a shape of blackness in the near distance.

The entire Tularosa Basin is tilted gently downhill, toward the southwest, like a great earthen flume. Elevations above sea level vary from around 6,000 feet at the northern end, to 5,400 near the town of Carrizozo, down to 4,000 feet at Lake Lucero. That incline is the simplest—and least mystical—explanation for what the lava did. Vomiting suddenly up from underground one stormy day, molten black rock flowed steaming and hissing off toward exactly that point from which came the White Sands—just as though some dark subterranean animus of alarming proportion were seeking to reunite itself, or maybe do battle, with its own antipodal twin.

Halfway there, the lava grew cool and viscid. Wrinkled with corrugations on its surface, pocked with gas bubbles, it slowed to a sloppy halt, congealing like a runnel of candlewax. It had traveled forty-four miles on a line, and lapped out across 120 square miles of desert. In some places it was a hundred feet deep.

The source of all this lava was a volcanic vent near the north end of the valley, a spigot-hole down to the Earth's liquid innards. The site of the vent is still distinguishable above the rest of the lava field by a high cone of cinder, a knoll known on the maps as Little Black Peak. Probably the lava poured out of this hole in two separate episodes, closely spaced. How long ago? The geologists can only guess. Maybe 2,000 years. Maybe less. Without question, the Carrizozo Malpais (a Spanish word for "badlands") is one of the youngest and best preserved lava fields in the continental U.S.

Its most striking characteristic is texture. The whole process of liquid rock flowing over rough terrain—cooling differentially from the outside in, piling up on itself into ropy corrugations and eddies, trapping those gas bubbles under thin-lidded domes—is captured as in a snapshot. Many of the big bubbles have collapsed to chasms, tiger pits thirty feet deep. Fissures have appeared. At some places the basalt, light and brittle stuff, has been broken to shards by the crowbar of weather. But in the scope of geologic time, weathering has hardly begun. The crowbar may have been briefly applied, but not yet the grinder, still less the emery cloth. This formation is jagged and raucous and therefore very new.

Some experts date the eruption to around 500 A.D. Archaeological evidence adds another interesting angle: Whenever the thing happened, apparently humans were in the valley to witness it.

They seem to have been pueblo-building people, sedentary agriculturalists, probably members of what is now referred to as the Mogollon Indian culture. Eventually they disappeared, or were driven out, to be replaced by ancestors of the Mescalero Apaches, a much different sort of bunch. No one knows just why the Mogollon folk went away. Possibly their departure was related to gradual changes in climate and water supply that were incompatible with their farming practices. Or it might have been war. Or something else.

One early commentator, writing in *Science* back in 1885, offered this: "A stream of no mean size seems to have once run down this valley. Not only has it now disappeared, but its bed is covered by lava and loose soil sometimes to great depths. As to the cause of the disappearance, it may have some connection with a tradition of the Indians which tells of a year of fire, when this valley was so filled with flames and poisonous gases as to be made uninhabitable."

In 1966 the state of New Mexico set aside a small tract of the lava flow for public enjoyment and edification. A parking lot and a set of restrooms have been added, also a short loop trail out through the Malpais, complete with number-keyed features of geologic and botanical interest for which commentary is supplied in a printed brochure. The place is fascinating, but largely unappreciated. To most people who visit it—and there aren't many—it is just a rest stop on the godforsaken two-lane between Carrizozo and another sleepy town. It is called, quite aptly, Valley of Fires State Park.

Back down at White Sands, public enjoyment and edification are overseen by the U.S. Interior Department. Early in 1933 (it seems to have been one of Herbert Hoover's last official acts) a presidential proclamation was signed, establishing White Sands National Monument. The borders of the Monument encompass a sizable swatch of the dune field, though by no means all of it. Lake Lucero is included, and enough area to provide a good sampling of the different dune types and the biotic communities that exist among them, but the lead edge of the dune field is far outside the Monument's northern boundary. At large, beyond Park Service jurisdiction; off on its own reckless chase.

And surrounding these two modest administrative units— literally arching over them through the sky—is another official fiefdom, one whose appointed mission does not include edifying the public, except perhaps in the most indirect way: White Sands Missile Range. Run by the Army, it is America's largest land-area shooting gallery for the testing of aeronautical and ballistic weaponry. Forty miles wide, stretching north and south for a hundred, it overlaps the whole Tularosa Basin almost exactly. Someone decided, back in 1945, that this is what the Tularosa was good for. *Hey, let's use that big piece of empty desert on the east side of the San Andres Mountains. It's perfect. And who will ever miss it?* Virtually no one has. Outsiders don't often come here to do their communing with God or nature. And the residents of Alamogordo, largest town in the Basin, seem generally to welcome the military dollars.

Today high-altitude research rockets go up and come down over the Missile Range. Cruise missiles under development by the Navy and the Air Force are launched from bombers, finding their way with magical sentience to targets out among the creosote bush. Drone aircraft—large jet-powered skeet, guided remotely— are blasted out of the sky by the latest and best in air-defense missiles, small scraps of their debris fluttering down onto the white dunes like exotic-alloy confetti. A laser-guided artillery projectile is fired off toward another corner of desert. From Fort Bliss, down near El Paso, Pershing II missiles fly up here on their intermediate-range trajectories, sailing over White Sands as though it were Poland, heading on toward the Malpais. Ground-to-ground missiles, air-to-air, air-to-ground and vice versa, every combination a country could need or believe it needs, the names themselves resonant with mythology and mystery and stouthearted martial precision: Nike, Talos, Tomahawk, Lance, Copperhead, Patriot, Stinger.

From one point of view, it is not so doleful as it might seem. The White Sands is a particularly delicate ecosystem where the margin of survival is narrow, for both animals and plants, and disturbances are not easily repaired. The entire chain of life there depends on a very slow process of soil formation and preliminary vegetal growth that occurs only in flat lowland areas between the active dunes. The slightest vehicle traffic (even heavy foot traffic) leaves scars across those flats that remain visible for decades, and an abusive degree of traffic could pull the bottom out from under the whole biotic community. Park Service regulations prohibit off-road vehicular traffic within the Monument, but that only accounts for a fraction of the dune field. Which is why Dr. William Reid, an ecologist from the University of Texas at El Paso who knows the White Sands as well as anyone, says: "Some people resent the Missile Range. I think it's a great boon in disguise. It's what saves the dunes and the inter-dune areas from the people in four-by-fours." A fragile environment gets Army protection from demented Sunday dirt-racers, and the occasional flaming rocket crash is—true enough—a small ecological price to pay.

So the innermost precincts of the Tularosa Basin remain unapproachable. You do not wander at will onto the White Sands Missile Range, either by truck or on foot. Fences and brusque signs warn you back. Electronic border sensors notice you. Polite security officers appear carrying shotguns. The message is *Keep out. This is secret stuff and we're busy. Besides you could get hurt.* Then, one day each year, the gate on a certain road swings open, and hundreds of vehicles drive through.

This annual motorcade begins from the parking lot of a K-Mart on the north edge of the town of Alamogordo, where the cars and the pickups and the RVs with out-of-state plates have gathered and pulled into file by 7:30 A.M. of the appointed day. It is the first Saturday in October, warm and bright, indecently good weather for a harvest picnic or a football game, and though neither of those are the purpose today, still the atmosphere is just a bit festive. A few children have been dragged along for the occasion, by parents or grandparents, and the kids burn off their energy dodging between bumpers, just as they would amid any dull gathering of stalled cars, for a fair or a funeral. Some folks have brought hampers of food; those in the open-topped sports cars wear jaunty hats. History is the sole attraction, but most of the people here assembled have come out in order to feel good about the particular moment of history in question. A much smaller segment—and they will be distinguishable when the picket signs appear—have come out in order to feel guilty and worried and bad. Only a few of us have come out, fully premeditated, in order to feel confused and ambivalent.

The men from the Alamogordo Chamber of Commerce (co-sponsors of the day's tour, along with the Missile Range management) do their directing of traffic with brisk, cheerful authority. After a few years' practice, they feel they have this thing down to a stroll. It is the Alamogordo equivalent of a big pancake breakfast, or a sweet-pea festival, or a rattlesnake roundup: both source and focus of civic pride. And sure enough, 250 autos move off right on schedule. The Missile Range fellows are holding their fire, but only so long. Just six hours have been set aside for these pilgrims to drive out across the Tularosa, deep into Missile Range property, on a thin asphalt road running between the White Sands and the black lava, to a place called Trinity Site; see the marker there, hear the speeches; and then get themselves back out of the line of fire.

Of course Trinity Site is the patch of woebegone desert from which Alamogordo, by metonymic incrimination, draws its greatest fame. Actually the site itself is sixty miles northwest of Alamogordo, shielded behind a stark upthrust of rock known as Oscura Peak. Here, on the morning of July 16, 1945, the nuclear age dawned gaudily. Robert Oppenheimer and his coven of young wizard physicists, from up in Los Alamos, had chosen the spot because of its sheer desolateness as a good place to test what was then still considered a dubious, improbable gadget.

The test was called Trinity. That code name had been supplied by Oppenheimer, from some free-associative inspiration about which he was ever afterward vague. In a letter, years later, he wrote: "Why I chose the name is not clear, but I know what thoughts were in my mind. There is a poem of John Donne, written just before his death, which I know and love. From it a quotation:

> . . . As West and East
> In all flatt Maps—and I am one—are one,
> So death doth touch the Resurrection.

That still does not make Trinity; but in another, better known devotional poem Donne opens, 'Batter my heart, three person'd God.' Beyond this, I have no clues whatever." For Robert Oppenheimer, an explanation like that was utterly characteristic.

Another Los Alamos scientist was directly in charge of the Trinity operation, but the two senior officials present were Oppenheimer and General Leslie Groves. Groves was a portly career soldier from the Army Corps of Engineers, a man with a large ego and an abrasive personality who had been thwarted in his hope of seeing overseas duty during World War II, and wound up instead, to his dismay, in command of the Manhattan Project. Actually Groves's original mandate seems to have been limited, a simple engineering task in the Corps tradition: to build the laboratories in which others would build the Bomb. But by degrees, in the wartime confusion, he filled a vacuum to become supreme potentate of the entire effort. It was Groves who had picked Robert Oppenheimer to run Los Alamos, the final-stage lab where the details of the weapon's design were worked out— picked him despite advice from the FBI that Oppenheimer was a security risk. Groves chose to ignore the FBI charges, but not without letting Oppenheimer know he had heard them. Which gave the general a certain leverage. Despite (or perhaps partly because of) that leverage, the two men had settled into a harmonious and effective working relationship.

In character, background, capabilities, these two couldn't have been more dissimilar. Oppenheimer was an intellectual of broad interests and surprisingly disparate eruditions, who read the classics of Greek and Sanskrit and Spanish literature, loved poetry, carefully studied the work of Karl Marx to see for himself what was there. He had come out of Harvard and done graduate fellowships at some of the best universities of Europe. He was an epicure. During the thirties he had been active in leftist causes, generous financially and with his time, never quite a card-carrying Communist but sympathetic with much that the Party was doing. Leslie Groves was an engineer and a soldier, period. Son of an austere Presbyterian minister who was himself also an Army man, a chaplain to the Fourteenth Infantry, Groves grew up on military posts and went to West Point. Like his father, he was something of a martinet. He was bullish and direct-minded and good at pushing straightforward jobs to completion; not so good at dealing with people. Impatient with psychological complexity. He knew nothing at all about nuclear physics when he was picked to ramrod the A-bomb project, but never allowed that to dampen his confidence in his own authority. Sometimes he was obtuse; sometimes, in the view of certain scientists, he behaved like a boob. Groves's military deputy on the project said later: "He's the biggest sonovabitch I've ever met in my life, but also one of the most capable individuals. . . . I hated his guts and so did everybody else but we had our form of understanding." Groves was a large man, well upholstered in flesh. Robert Oppenheimer was gangly and emaciated. Both of them could be arrogant, both could be quick to judge. By objective criteria, they should have been expected to loathe and distrust each other wholeheartedly. But it didn't unfold that way. As an unlikely partnership, the physicist and the soldier also evidently had their own form of understanding.

And so Trinity happened, a great success.

Originally the test firing was scheduled for 4:00 A.M. on the morning of July 16, a Monday. By Saturday afternoon the bomb had been assembled on the site—its plutonium core inserted delicately into the larger casing—and hoisted up to the top of its 100-foot girder tower. Electrical detonators were connected at sixty-four points along the outside of the metal sphere, small taps plugged onto the surface, a tangle of crisscrossing wires, as though the monstrous and inscrutable thing were having its mind read by electroencephalograph. What was it thinking? What did it know? Notwithstanding the elaborate electronics, that could only be found out the hard way. Before sundown on the last afternoon Robert Oppenheimer himself climbed the tower, alone, for a last look at this device he had guided into being. About the same time, General Groves arrived at the site. Then the weather turned bad.

Throughout Sunday night the Trinity gadget sat atop its steel tower in the midst of a desert storm, a raucous overture of thunder and lightning and wind-driven rain. No one seems to have gotten an hour's sleep except Leslie Groves. Oppenheimer paced and fretted. One bolt of lightning striking the tower might not have detonated the bomb, but it certainly would have destroyed the electrical circuitry and caused a major delay. Any delay now was dreaded, because Harry Truman at Potsdam was eager for news about this far-fetched atomic weapon, and the test results would tell him how to deal with Stalin concerning the continuing war against Japan. But lightning wasn't the only meteorological problem. There was also concern that storm clouds would carry large doses of fallout onto population centers downwind. Amarillo 300 miles away and, more immediately, the small town of Carrizozo, just east across the Tularosa Basin. Groves woke from his nap and consulted with Oppenheimer.

Then around 4:00 A.M. the rain stopped. The countdown resumed. A young scientist named Joe McKibben, responsible for the remote electrical signals, threw a switch at minus 45 seconds that locked the whole system into an automatic timer. Though not unstoppable, the event was now progressing on its own impetus, without active human control. In the command bunker and back at the base camp, people lay down on their bellies with their feet toward the tower, a position of backward obeisance. At 5:29:45 A.M. Mountain War Time, the wrapper of high explosives squeezed down on the plutonium core. Neutrons ricocheted, atoms split; a chain reaction ensued.

The report which General Groves sent off at once to Potsdam said: "For a brief period there was a lighting effect within a radius of 20 miles equal to several suns at midday; a huge ball of fire was formed which lasted for several seconds. This ball mushroomed and rose to a height of over 10,000 feet before it dimmed. The light from the explosion was seen clearly at Albuquerque, Santa Fe, Silver City, El Paso, and other points generally to about 180 miles away. . . . A massive cloud was formed which surged and billowed upward with tremendous power, reaching the substratosphere at an elevation of 41,000 feet, 36,000 feet above the ground, in about 5 minutes, breaking without interruption through a temperature inversion at 17,000 feet which most of the scientists thought would stop it. . . . Huge concentrations of highly radioactive materials resulted from the fission and were contained in this cloud." That version reached Truman by courier. In a quiet moment sometime afterward, Robert Oppenheimer offered his own version: "A few people laughed, a few people cried. Most people were silent." Then he quoted another bit of poetry.

The command bunker and the base camp are long since gone. Oppenheimer and Leslie Groves are gone. Hiroshima and Nagasaki are not the same cities they were. But Trinity Site was just a spot out in the desert, and so it remains.

On the first Saturday of October, four decades later, you can still see the frizzled steel stumps left behind when the tower was vaporized. You can still pick up a chirpy reading on a Geiger counter. You can hear speeches by the commander of the Missile Range, by the president of the Chamber of Commerce, and you can join in prayer with an Army chaplain who says ". . . and guide us, Lord, that we may then begin beating our plowshares into . . . uh, beating our swords into plowshares." The confusion is understandable. You can even chat with Joe McKibben, the man who threw that last switch, now a retirement-age gentleman in casual clothes with a friendly and slightly dotty manner, who happens to be back for the tour this year himself. McKibben is genial about answering questions but there is an unreachable look in his eyes.

He says: "Well, you have to wonder how it would have gone if some things had been different."

There is one more stop on the Tularosa circuit; one more element in the Tularosa design. This one could easily be overlooked, so watch carefully for a small sign along the two-lane that runs north out of Alamogordo toward Carrizozo. Again you will be about halfway from White Sands to the Malpais, but no motorcade surrounding you now, no men waving you forward with flashlights, no guardhouses lifting their crossbars. You stay alert for an inconspicuous junction, at the corner of which sits the bulky white shape of what once was a roadhouse. The roadhouse is boarded up; lettering on its window says *3 Rivers and Deviants M/C Clubhouse. Private. Beware of dog*. That's the turn. You drive east for another five miles on dirt, into the foothills of the Sacramento range. You park and begin walking, up the crest of a sharp north-south ridge. Another sign shows you the way. Welcome to the Three Rivers Petroglyph Site, famous among archaeologists of the Southwest and unknown to almost everyone else.

A community of the Mogollon people, those pacific agriculturalists, lived here at the start of the present millennium for a span of about 400 years. Their village was down lower in the Three Rivers drainage; this exposed ridge seems to have served as a lookout, from which they could spot game or approaching enemies far out in the Tularosa Basin. The view is indeed good. Gazing westward across the desert, you can see the great white shape of gypsum and the great black shape of lava. You can see the barren upthrust that is Oscura Peak and, if a fireball 10,000 feet high were to suddenly blossom behind it, you would sure as hell see that too. Mogollon scouts may have spent many hours and weeks and years up on this boulder-toothed ridge, watching. After four centuries, though, the whole community disappeared.

Probably they migrated north, out of the Tularosa. They could have been fleeing a drought. Or it might have had some link with that dolorous tradition, the one which tells of "a year of fire, when this valley was so filled with flames and poisonous gases as to be made uninhabitable." Maybe they saw something disturbing. A sudden thunderous ebullition of cinder and smoke and liquidy black rock, for instance. Or who knows what.

All they left us were a few potsherds, a few fallen adobe pit-houses, and about 5,000 rock carvings, scratched and chipped onto the boulder faces along that ridge.

Some of these carvings are far more artful than others. Some are vividly representational—a bighorn sheep impaled by three arrows—and some are abstract. There are a few human figures, but no dashing romantic portrayals of prowess in battle or hunting; mainly large ovoid heads, wide-eyed and jug-eared. No warriors on horseback. Aside from those arrows in the bighorn, weaponry is conspicuously absent. Animal portraits abound, especially birds and horned mammals and even a few fish. Also there are carven footprints: bear paws, bird prints as though in mud, human foot shapes. A majority of the petroglyphs on the ridge, though, are abstract designs.

Among these the most common motif, appearing in many variations, is a circle or several concentric circles surrounded by a ring of dots. Similar circle-and-dot patterns are known from Mogollon petroglyph sites throughout the Southwest, but they seem to have held a particular fascination for the artists at Three Rivers. You can see in them almost anything you might choose: a circle of family members, the solar system, the nucleus and electrons of an atom.

Walking the ridge trail up among these carvings, you find another arresting motif. Having made a lucky detour off the main path, watching for rattlesnakes as you step, you notice it first on the western side of a large dark boulder cropping out high on the ridge's westernmost knob. This design is more elaborate and sophisticated than the others, even beautiful, and something about it stops you short:

It recurs, in its variations, a dozen times on the ridge. In a few instances it is less squarish, more curvilinear; in several it is done with linked or interlocking spirals. Despite transmutations, in each the essence is unmistakable.

Evidently the people at Three Rivers had a concept of yin and yang. They drew rock pictures of dialectical oneness. They cherished some notion—maybe it was only wishful—of a unity within which all worldly flux remains balanced.

What did that mean to them? Obviously we don't know. They watched and they drew and then they departed. But the yin-and-yang concept, after all, is easily enough reducible to truism; and these figures, like the dotted circles, can be taken to represent almost anything. John Donne's idea, for instance: "As West and East . . . are one,/So death doth touch the Resurrection." Or another idea, maybe in this case more applicable: "For they have sown the wind, and they shall reap the whirlwind."

Now suddenly a pair of turkey vultures wheel into view above you, cruising on thermals that rise off the west slope of the Three Rivers ridge. With typical lazy grace, they are scouting for a meal. One of the vultures sweeps closer to scrutinize you.

This bird pauses, holding position not thirty feet over your head, like a kite on a short string. It seems unsure whether to take you for a pile of dead meat. And you *are* sitting quite still. The confusion is understandable.

# Partial Sources

Magazine formats generally do not allow space for bibliographies, nor encourage footnoting. Consequently the information, in a magazine article or column, seems sometimes to come from nowhere. To rectify that illusion at least partly, in the case of these pieces, I list here some (but not all) of the printed and manuscript sources from which I have cannibalized facts and reaped understanding. There are still other sources to which I'm indebted—many, more than I can list, more than I can remember. Interviews have also been helpful, though less often than libraries. And occasionally, of course, I just make things up.

This list might also be of use to any reader who wants to chase a particular subject further.

## SEA CUCUMBERS

*Grzimek's Animal Life Encyclopedia.* Bernhard Grzimek, editor-in-chief. New York: Van Nostrand Reinhold Co. 1972. (The entire Grzimek encyclopedia, volumes I through XIII, was useful to me on a number of these essays.)

# BATS

*Bats of America.* Roger W. Barbour and Wayne H. Davis. Lexington, Ky.: The University Press of Kentucky. 1969.

*The Lives of Bats.* D.W. Yalden and P.A. Morris. New York: Quadrangle/ The New York Times Book Co. 1975.

*Bats.* Glover Morrill Allen. New York: Dover Publications. 1962.

"Bats Away!" Joe Michael Feist. In *American Heritage*, April-May, 1982.

*Biology of Bats.* Vols. I & II. Edited by William A. Wimsatt. New York: Academic Press. 1970.

*Silently, By Night.* Russell Peterson. New York: McGraw-Hill Book Co. 1964.

*Bat Research News* (originally *Bat Banding News*). Vols. 1-16: 1960-1975. Compiled by Wayne H. Davis (Department of Zoology, University of Kentucky, Lexington, Ky.) from founding to April, 1970; compiled by Robert L. Martin (Department of Biology, University of Maine, Farmington, Maine) thereafter.

# BLACK WIDOWS

*The Black Widow Spider.* Raymond W. Thorp and Weldon D. Woodson. New York: Dover Publications. 1976.

"On the Great Abundance of the Black Widow Spider." Albert Milzer. In *Science*; November 2, 1934.

# OCTOPI

*Octopus: Physiology and Behaviour of an Advanced Invertebrate.* M.J. Wells. London: Chapman and Hall. 1978.

*Kingdom of the Octopus: The Life-History of the Cephalopoda.* Frank W. Lane. London: Jarrolds Publishers Ltd. 1957.

*Octopus and Squid: The Soft Intelligence.* Jacques-Yves Cousteau and Philippe Diolé. Translated by J.F. Bernard. New York: Doubleday and Co. 1973.

"The Giant Pacific Octopus." William L. High. In *Marine Fisheries Review*; September, 1976.

"Cephalopods Do It Differently." Martin Wells. In *New Scientist*; November 3, 1983.

# MOSQUITOES

*The Mosquito: Its Life, Activities, and Impact on Human Affairs.* J.D. Gillett. New York: Doubleday and Co. 1972.

*The Natural History of Mosquitoes.* Marston Bates. New York: The Macmillan Co. 1949.

*Mosquitoes: Their Bionomics and Relation to Disease.* William R. Horsfall. New York: Hafner Publishing Co. 1972.

*Mosquitoes, Malaria and Man: A History of the Hostilities Since 1880.* Gordon Harrison. New York: E.P. Dutton. 1978.

*Conversion of Tropical Moist Forests.* A report to the Committee on Research Priorities in Tropical Biology of the National Research Council. Norman Myers. Washington, D.C.: National Academy of Sciences. 1980.

"Eco-crime on the Equator." Janet Marinelli. In *Environmental Action*; March, 1980. Published by Environmental Action, Inc., Washington, D.C.

# CROWS

*Crows, Jays, Ravens and Their Relatives.* Sylvia Bruce Wilmore. London: David and Charles. 1977.

*The Crows: A Study of the Corvids of Europe.* Franklin Coombs. London: B.T. Batsford Ltd. 1978.

*Crows of the World.* Derek Goodwin. Ithaca, N.Y.: Comstock Publishing Associates (a division of Cornell University Press) in cooperation with the British Museum (Natural History). 1976.

*Ravens, Crows, Magpies, and Jays.* Tony Angell. Seattle: University of Washington Press. 1978.

"Anting and the Problem of Self-stimulation." K.E.L. Simmons. In *Journal of Zoology*, vol. 149. 1966.

"A Review of the Anting-Behaviour of Passerine Birds." K.E.L. Simmons. In *British Birds*, vol. L. October, 1957.

"Avian Play." Millicent S. Ficken. In *The Auk*, vol. 94. July, 1977.

# VOX POPULI

*Knowledge, Affection, and Basic Attitudes Toward Animals in American Society—Phase 3.* Stephen R. Kellert and Joyce K. Berry. Results of a study funded by the U.S. Fish and Wildlife Service. Springfield, Va.: National Technical Information Service. 1980.

## ANACONDAS

*The Giant Snakes.* Clifford H. Pope. London: Routledge and Kegan Paul. 1962.
*Giant Reptiles.* Sherman A. Minton, Jr. and Madge Rutherford Minton. New York: Charles Scribner's Sons. 1973.
*On the Track of Unknown Animals.* Bernard Heuvelmans. Translated and abridged by Richard Garnett. New York: Hill and Wang. 1965.

## MOTHS

"Blood-sucking Moths of Malaya." Hans Bänziger. In *Fauna,* 1971.
"Preliminary Observations on a Skin-Piercing Blood-Sucking Moth (*Calyptra eustrigata* (Hymps.) (Lep., Noctuidae)) in Malaya." H. Bänziger. In *Bulletin of Entomological Research,* vol. 58. 1968.
"Records of Eye-frequenting Lepidoptera from Man." H. Bänziger and W. Buttiker. In *Journal of Medical Entomology,* vol. 6. January 30, 1969.
"Blood-feeding Habits of Adult Noctuidae (Lepidoptera) in Cambodia." W. Buttiker. In *Nature,* vol. 184. October 10, 1959.
*The Illustrated Encyclopedia of the Butterfly World.* Paul Smart. London: The Hamlyn Publishing Group Ltd. 1976.
*Butterflies.* Photographs by Kjell B. Sandved. Text by Jo Brewer. New York: Harry N. Abrams, Inc. 1976.
*An Introduction to the Study of Insects.* Donald J. Borror and Dwight M. DeLong. New York: Holt, Rinehart and Winston, Inc. 1971. (Borror and DeLong, like the Grzimek encyclopedia, has been generally quite valuable to me as a reference work.)

## COCKROACHES

*The Structure and Life-History of the Cockroach* (Periplaneta orientalis). L.C. Miall and Alfred Denny. London: L. Reeve and Co. 1886.
*The Cockroach.* P.B. Cornwell. Hutchinson of London. 1968.

*The Biology of the Cockroach.* D.M. Guthrie and A.R. Tindall. New York: St. Martin's Press. 1968.

"The Effect of Radiation on the Longevity of the Cockroach, *Periplaneta americana*, as Affected by Dose, Age, Sex and Food Intake." D.R.A. Wharton and Martha L. Wharton. In *Radiation Research*, vol. 11. 1959.

"The Life History of the American Cockroach, *Periplaneta americana* Linn. (Orthop.: Blattidae)." Phil Rau. In *Entomological News*, vol. LI. 1940.

*The Fate of the Earth.* Jonathan Schell. New York: Alfred A. Knopf. 1982.

## JACK HORNER

"The Nesting Behavior of Dinosaurs." John R. Horner. In *Scientific American*; April, 1984.

"Evidence of Colonial Nesting and 'Site Fidelity' among Ornithischian Dinosaurs." John R. Horner. In *Nature*, vol. 297. June 24, 1982.

"Nest of Juveniles Provides Evidence of Family Structure Among Dinosaurs." John R. Horner and Robert Makela. In *Nature*, vol. 282. November 15, 1979.

"Anatomical and Ecological Evidence of Endothermy in Dinosaurs." Robert T. Bakker. In *Nature*, vol. 238. July 14, 1972.

"Dinosaur Renaissance." Robert T. Bakker. In *Scientific American*, vol. 232. April, 1975.

"Terrestrial Vertebrates as Indicators of Mesozoic Climates." John H. Ostrom. In *Proceedings of the North American Paleontological Convention.* 1969.

*The Hot-Blooded Dinosaurs.* Adrian J. Desmond. New York: The Dial Press/ James Wade. 1975.

## EUGÈNE MARAIS

*The Soul of the White Ant.* Eugène N. Marais. With a biographical note by his son. Translated by Winifred de Kok. Harmondsworth: Penguin Books. 1973.

*The Soul of the Ape.* Eugène N. Marais. With an introduction by Robert Ardrey. Harmondsworth: Penguin Books. 1973.

*African Genesis: A Personal Investigation into the Animal Origins and Nature of Man.* Robert Ardrey. New York: Atheneum. 1967.

# TYCHO BRAHE

*Tycho Brahe.* J.L.E. Dreyer. Edinburgh: Adam and Charles Black. 1890.
*The Sleepwalkers: A History of Man's Changing Vision of the Universe.* Arthur Koestler. New York: Grosset and Dunlap. 1963.
*The New Astronomy.* Paul and Leslie Murdin. New York: Thomas Y. Crowell Co. 1978.
*The Practical Astronomer.* Colin A. Ronan. London: Roxby Press Ltd. 1981.

# CRICK AND OTHERS

*Worlds in the Making: The Evolution of the Universe.* Svante Arrhenius. New York: Harper and Brothers. 1908.
*The Origin of Life on the Earth.* A.I. Oparin. Translated by Ann Synge. New York: Academic Press. 1957.
*The Origin of Life.* J.D. Bernal. New York and Cleveland: The World Publishing Co. 1967.
*Life Itself: Its Origin and Nature.* Francis Crick. New York: Simon and Schuster. 1981.
"Directed Panspermia." F.H.C. Crick and L.E. Orgel. In *Icarus*, vol. 19. 1973.
"The Zoo Hypothesis." John A. Ball. In *Icarus*, vol. 19. 1973.

# TREE PEOPLE

"An Ancient Bristlecone Pine Stand in Eastern Nevada." Donald R. Currey. In *Ecology*, vol. 46. Early Summer, 1965.
"Bristlecone Pine: Science and Aesthetics." C.W. Ferguson. In *Science*, vol. 159. February 23, 1968.
"Environment in Relation to Age of Bristlecone Pines." Valmore C. LaMarche, Jr. In *Ecology*, vol. 50. Winter, 1969.
"Recognizing Site Adversity and Drought-Sensitive Trees in Stands of Bristlecone Pine (*Pinus longaeva*)." R.S. Beasley and J.O. Klemmedson. In *Economic Botany*, vol. 27. January-March, 1973.

*The Tree*. John Fowles and (photographs by) Frank Horvat. Boston: Little, Brown and Co. 1979.

*The Tree Book*. Julia Ellen Rogers. New York: Doubleday, Page and Co. 1922.

*Trees of America*. By the editors of *Outdoor World*. Waukesha, WI: Country Beautiful Corp. 1973.

# GRAYLING

"Life History and Management of the American Grayling (*Thymallus signifer tricolor*) in Montana." Perry H. Nelson. In *Journal of Wildlife Management*, vol. 18. July, 1954.

"Grayling of Grebe Lake, Yellowstone National Park, Wyo." Thomas E. Kruse. In *Fishery Bulletin of the Fish and Wildlife Service*, vol. 59. 1958.

"Montana Grayling and Its Habitat." E. Earl Willard and Margaret Herman. A pamphlet (in typescript), sponsored by the U.S. Forest Service.

"Montana Grayling: The Lady of the Streams." George D. Holton. In *Montana Outdoors*; September-October, 1971. For facts and numbers concerning the hatchery-and-planting program in Montana, I am also indebted to George Holton (personal communication), and to a fact-sheet prepared by Bill Gould. Neither man is accountable for the conclusions I've drawn.

# TIGERS

*The Face of the Tiger*. Charles McDougal. London: Rivington Books and Andre Deutsch. 1977.

*The Deer and the Tiger: A Study of Wildlife in India*. George B. Schaller. Chicago: University of Chicago Press. 1967.

*Saving the Tiger*. Guy Mountfort. New York: The Viking Press. 1981.

*Tiger!: The Story of the Indian Tiger*. Kailash Sankhala. New York: Simon and Schuster. 1977.

*The Social Organization of Tigers* (Panthera tigris) *in Royal Chitawan National Park, Nepal*. Melvin E. Sunquist. Smithsonian Contributions to Zoology, No. 336. Washington, D.C.: Smithsonian Institution Press. 1981.

"Some Observations on Tiger Behaviour in the Context of Baiting." Charles McDougal. In *Journal of the Bombay Natural History Society*, vol. 77. April 27, 1981.

## CONDORS

*The California Condor*. C.B. Koford. Research Report No. 4 of the National Audubon Society. New York. 1953.

*The Current Status and Welfare of the California Condor*. Alden H. Miller, Ian I. McMillan, and Eben McMillan. Research Report No. 6 of the National Audubon Society. New York. 1965.

"The Death of William Faulkner." Hughes Rudd. In *My Escape from the CIA and Other Improbable Events*. New York: E.P. Dutton. 1966.

*Man and the California Condor: The Embattled History and Uncertain Future of North America's Largest Free-living Bird*. Ian McMillan. New York: E.P. Dutton and Co. 1968.

*The California Condor, 1966-76: A Look at Its Past and Future*. Sanford R. Wilbur, U.S. Fish and Wildlife Service. In *North American Fauna*, vol. 72.

"Last Days of the Condor?" Faith McNulty. A two-part article in *Audubon*, 1978.

*The Condor Question: Captive or Forever Free?* Edited by David Phillips and Hugh Nash. San Francisco: Friends of the Earth. 1981.

## BISON

"History of the Bison in Yellowstone Park." Curtis K. Skinner and Wayne B. Alcorn. Typescript, in the archives of the Research Library, Yellowstone National Park. 1941 and (updated) 1942-51.

"Preliminary Report on the Hayden Valley Bison Range." Walter H. Kittams. Typescript, in the archives of the Research Library, Yellowstone National Park. 1949.

*The Bison of Yellowstone National Park*. Mary Meagher. Washington, D.C.: National Park Service. 1973.

*North American Bison: Their Classification and Evolution*. Jerry N. McDonald. Berkeley: University of California Press. 1981.

*The Buffalo*. Francis Haines. New York: Thomas Y. Crowell Co. 1970.

*The Buffalo Book: The Full Saga of the American Animal*. David A. Dary. Chicago: The Swallow Press, Inc. 1974.


## ANIMAL RIGHTS

*Animal Liberation: A New Ethics for Our Treatment of Animals*. Peter Singer. New York: A New York Review Book. 1975.

*The Case for Animal Rights*. Tom Regan. Berkeley: University of California Press. 1983.

*Animal Rights and Human Obligations*. Edited by Tom Regan and Peter Singer. Englewood Cliffs, N.J.: Prentice-Hall. 1976.

*All That Dwell Therein: Animal Rights and Environmental Ethics*. Tom Regan. Berkeley: University of California Press. 1982.

*On the Fifth Day: Animal Rights and Human Ethics*. Edited by Richard Knowles Morris and Michael W. Fox. Washington, D.C.: Acropolis Books, Ltd. 1978.

"Renegotiating the Contracts." Barry Lopez. In *Parabola*; Spring, 1983.


## EXTINCTION

*The Sinking Ark: A New Look at the Problem of Disappearing Species*. Norman Myers. Oxford: Pergamon Press. 1979.

*Extinction: The Causes and Consequences of the Disappearance of Species*. Paul and Anne Ehrlich. New York: Random House. 1981.

*Disappearing Species: The Social Challenge*. Erik Eckholm. Worldwatch Paper 22. From the Worldwatch Institute, Washington, D.C. 1978.

*Wildlife in America*. Peter Matthiessen. Harmondsworth: Penguin Books. 1977.


## SEMELPARITY

"Selection for Optimal Life Histories: The Effects of Age Structure." William M. Schaffer. In *Ecology*, vol. 55. Early Spring, 1974.

"The Adaptive Significance of Variations in Reproductive Habit in the Agavaceae II: Pollinator Foraging Behavior and Selection for Increased Reproductive Expenditure." William M. Schaffer and M. Valentine Schaffer. In *Ecology*, vol. 60. October, 1979.

"Selection for Optimal Life Histories. II: Multiple Equilibria and the Evolution of Alternative Reproductive Strategies." William M. Schaffer and Michael L. Rosenzweig. In *Ecology*, vol. 58. Winter, 1977.

"Why Bamboos Wait So Long to Flower." Daniel H. Janzen. In *Annual Review of Ecology and Systematics*, vol. 7. 1976.

"Life Historical Consequences of Natural Selection." Madhav Gadgil and William H. Bossert. In *The American Naturalist*, vol. 104. January-February, 1970.

"Life-History Tactics: A Review of the Ideas." Stephen C. Stearns. In *The Quarterly Review of Biology*, vol. 51. March, 1976.

*The Salmon: Their Fight for Survival*. Anthony Netboy. Boston: Houghton Mifflin. 1974.

*Pacific Salmon and Steelhead Trout*. R.J. Childerhose and Marj Trim. Seattle: University of Washington Press. 1979.

*Agaves of Continental North America*. Howard Scott Gentry. Tucson: University of Arizona Press. 1982.

## LIVING WATER

*The Ecology of Running Waters*. H.B.N. Hynes. Toronto: University of Toronto Press. 1970.

## HYPOTHERMIA

I benefited from—in addition to the Ted Lathrop pamphlet and the Associated Press story—the help of Mark Smith, a reporter for the *Tri-County Tribune* of Deer Park, Washington, not far from Chattaroy. Mr. Smith shared with me his coverage of the Ram Patrol misfortune.

## PARTHENOGENESIS

"Parthenogenesis in Rotifers: The Control of Sexual and Asexual Reproduction." C. William Birky, Jr. and John J. Gilbert. In

*The American Zoologist*, vol. 11. 1971. (This volume of *The American Zoologist* contains a whole series of papers on the subject, by a dozen different authors; each of the papers had been presented at a symposium on parthenogenesis, part of the meetings of the American Association for the Advancement of Science, in Chicago, December, 1970.)

*Animal Cytology and Evolution*. M.J.D. White. Cambridge, England: Cambridge University Press. 1954.

*The Economy of Nature and the Evolution of Sex*. Michael T. Ghiselin. Berkeley: University of California Press. 1974.

*Aphids*. Roger Blackmun. London: Ginn and Company Ltd. 1974.

*Perspectives in Aphid Biology*. Edited by A.D. Lowe. Auckland, New Zealand: The Entomological Society of New Zealand (Inc.). 1973.

*Parthenogenesis and Polyploidy in Mammalian Development*. R.A. Beatty. Cambridge, England: Cambridge University Press. 1957.

"Significance of Parthenogenesis in the Evolution of Insects." Esko Suomalainen. In *Annual Review of Entomology*, vol. 7.

# DESERTS

*The North American Deserts*. Edmund C. Jaeger. Stanford: Stanford University Press. 1957.

*The Journey Home: Some Words in Defense of the American West*. Edward Abbey. New York: E.P. Dutton. 1977.

*Deserts*. Gayle Pickwell. New York: McGraw-Hill. 1939.

*The Deserts of Central Asia*. M.P. Petrov. Washington, D.C.: Joint Publications Research Service. 1966.

*In the Deserts of This Earth*. Uwe George. Translated by Richard and Clara Winston. New York: Harcourt Brace Jovanovich. 1977.

*The Voice of the Desert: A Naturalist's Interpretation*. Joseph Wood Krutch. New York: William Morrow and Co./ Morrow Quill Paperbacks. 1980.

# BLUBBER

*Arctic Life of Birds and Mammals*. Laurence Irving. New York: Springer-Verlag. 1972.

*The Behavior and Physiology of Pinnipeds.* Edited by R.J. Harrison, Charles E. Rice and others. New York: Appleton-Century-Crofts. 1968.
*Seals of the World.* Gavin Maxwell. Boston: Houghton Mifflin. 1967.
*Seals, Sea Lions, and Walruses: A Review* of Pinnipedia. Victor B. Scheffer. Stanford: Stanford University Press. 1958.

# TULAROSA

"Lost Rivers." M.W. Harrington. In *Science*, vol. VI. September 25, 1885.
"Hydrologic Control Over the Origin of Gypsum at Lake Lucero, White Sands National Monument, New Mexico." Roger J. Allmendinger. A Master of Science thesis, in typescript, in the library at the headquarters of White Sands National Monument.
*Final Report: White Sands National Monument Natural Resources and Ecosystem Analysis.* William H. Reid, project director. (Typescript.) The University of Texas at El Paso: Laboratory for Environmental Biology, Research Report Number 12. October, 1980.
"An Ecological Reconnaissance in the White Sands, New Mexico." Fred W. Emerson. In *Ecology*, vol. 16. April, 1935.
"Petroglyphs at Three Rivers, New Mexico: A Partial Survey." Kay Sutherland. In *The Artifact* (El Paso Archeological Society), vol. 16.
*Rock Art in New Mexico.* Polly Schaafsma. Albuquerque: University of New Mexico Press. 1975.
*Robert Oppenheimer: Letters and Recollections.* Edited by Alice Kimball Smith and Charles Weiner. Cambridge, Mass.: Harvard University Press. 1980.
*Now It Can Be Told: The Story of the Manhattan Project.* Leslie R. Groves. New York: Da Capo Press. 1975.
*Manhattan Project: The Untold Story of the Making of the Atomic Bomb.* Stephane Groueff. New York: Bantam. 1968.
*J. Robert Oppenheimer: Shatterer of Worlds.* Peter Goodchild. Boston: Houghton Mifflin. 1981.
The quotations in this essay come from the following:

"Why I chose the name . . ." is quoted from Smith and Weiner, p. 290.

"He's the biggest sonovabitch . . ." is quoted from Goodchild, pp. 56-57.

"For a brief period there was a lighting effect . . ." is quoted from Groves, pp. 433-434.

"A few people laughed, a few people cried . . ." is quoted from Goodchild, p. 162. Footage of this statement by Oppenheimer also appears in the excellent film documentary *The Day After Trinity*, distributed by Pyramid Film and Video, Santa Monica, California.